Fundamental Concepts of Biology

Fundamental Concepts of Biology

Nicola Galecki

New York

Published by Syrawood Publishing House,
750 Third Avenue, 9th Floor,
New York, NY 10017, USA
www.syrawoodpublishinghouse.com

Fundamental Concepts of Biology
Nicola Galecki

© 2020 Syrawood Publishing House

International Standard Book Number: 978-1-64740-015-6 (Hardback)

This book contains information obtained from authentic and highly regarded sources. All chapters are published with permission under the Creative Commons Attribution Share Alike License or equivalent. A wide variety of references are listed. Permissions and sources are indicated; for detailed attributions, please refer to the permissions page. Reasonable efforts have been made to publish reliable data and information, but the authors, editors and publisher cannot assume any responsibility for the validity of all materials or the consequences of their use.

Trademark Notice: Registered trademark of products or corporate names are used only for explanation and identification without intent to infringe.

Cataloging-in-Publication Data

Fundamental concepts of biology / Nicola Galecki.
   p. cm.
Includes bibliographical references and index.
ISBN 978-1-64740-015-6
1. Biology. 2. Life sciences. 3. Life (Biology). 4. Natural history.
I. Galecki, Nicola.
QH307.2 .F86 2020
570--dc23

TABLE OF CONTENTS

Preface .... **VII**

Chapter 1 **Biology: An Introduction** 1
- Branches of Biology 12

Chapter 2 **Cell: The Fundamental Unit of Life** 16
- Cell 16
- Cell Theory 68
- Cell Types 71
- Cell Functions 82
- Cellular Respiration 83
- Cellular Reproduction 87
- Cell Cycle 88

Chapter 3 **Reproduction in Organisms** 92
- Reproduction 92
- Reproduction in Plants 107
- Reproduction in Humans 112

Chapter 4 **Inheritance and Variation** 143
- Biological Inheritance 143
- Mendel's Experiment 144
- Laws of Inheritance 147
- Heredity 151
- Variation 161
- Mutation 163

Chapter 5 **Theories and Concepts of Evolution** 166
- Evolution 166
- Theories of Evolution 185
- Common Descent 199
- Speciation 202
- Coevolution 205
- Adaptation 207
- Adaptive Radiation 210

Permissions

Index

PREFACE

Biology is a branch of natural science which deals with the study of life and living organisms. It focuses on their physical structure, molecular interactions, physiological mechanisms, chemical processes, development and evolution. It considers cells as the fundamental unit of life and genes as the basic unit of heredity. It also perceives evolution as the force that is responsible for the creation and extinction of species. There are a number of sub-disciplines within biology such as theoretical biology and experimental biology. Theoretical biology uses mathematical methods for the formulation of quantitative models. Experimental biology conducts empirical experiments for testing the validity of proposed theories and understanding the mechanisms underlying life. This textbook is a valuable compilation of topics, ranging from the basic to the most complex theories and principles in the field of biology. Some of the diverse topics covered herein address the varied branches that fall under this category. Those in search of information to further their knowledge will be greatly assisted by this book.

A short introduction to every chapter is written below to provide an overview of the content of the book:

Chapter 1 - Biology is the branch of science that deals with the study of life and living organisms. It studies their chemical processes, physiological mechanisms, physical structure, evolution, etc. This is an introductory chapter which will introduce briefly all these aspects of biology.; **Chapter 2** - The cell is the fundamental and smallest unit of life of all known living organisms. There are two types of cells- eukaryotic and prokaryotic cells. Cellular respiration and cellular reproduction are the two main functions of cells. The chapter closely examines these key concepts related to cells to provide an extensive understanding of the subject.; **Chapter 3** - The biological process by which new individual organisms are produced from their parents is known as reproduction. Sexual and asexual are the two forms of reproduction. Plant reproduction consists of both sexual and asexual reproduction whereas human reproduction is a form of sexual reproduction. The topics elaborated in this chapter will help in gaining a better perspective about these types of reproduction as well as the processes related to them.; **Chapter 4** - Biological inheritance is the process of passing the traits from parents to their offspring. The offspring cells and organisms inherit the genetic information of their parents either by asexual reproduction or sexual reproduction. All these diverse principles of inheritance, mutation and variation have been carefully analyzed in this chapter.; **Chapter 5** - The change in genetic characteristics of biological population over consecutive generations is known as evolution. These characteristics are the expressions of genes that are passed on from parent to offspring during reproduction. Theories of evolution include common descent, coevolution, speciation, adaptation, etc. This chapter discusses in detail these theories and methodologies related to evolution.

Finally, I would like to thank my fellow scholars who gave constructive feedback and my family members who supported me at every step.

Nicola Galecki

Chapter 1
Biology: An Introduction

Biology is the branch of science that deals with the study of life and living organisms. It studies their chemical processes, physiological mechanisms, physical structure, evolution, etc. This is an introductory chapter which will introduce briefly all these aspects of biology.

Biology is the study of living things and their vital processes. The field deals with all the physico-chemical aspects of life. The modern tendency toward cross-disciplinary research and the unification of scientific knowledge and investigation from different fields has resulted in significant overlap of the field of biology with other scientific disciplines. Modern principles of other fields—chemistry, medicine, and physics, for example—are integrated with those of biology in areas such as biochemistry, biomedicine, and biophysics.

Biology is subdivided into separate branches for convenience of study, though all the subdivisions are interrelated by basic principles. Thus, while it is custom to separate the study of plants (botany) from that of animals (zoology), and the study of the structure of organisms (morphology) from that of function (physiology), all living things share in common certain biological phenomena—for example, various means of reproduction, cell division, and the transmission of genetic material.

Biology is often approached on the basis of levels that deal with fundamental units of life. At the level of molecular biology, for example, life is regarded as a manifestation of chemical and energy transformations that occur among the many chemical constituents that compose an organism. As a result of the development of increasingly powerful and precise laboratory instruments and techniques, it is possible to understand and define with high precision and accuracy not only the ultimate physiochemical organization (ultrastructure) of the molecules in living matter but also the way living matter reproduces at the molecular level. Especially crucial to those advances was the rise of genomics in the late 20th and early 21st centuries.

Cell biology is the study of cells—the fundamental units of structure and function in living organisms. Cells were first observed in the 17th century, when the compound microscope was invented. Before that time, the individual organism was studied as a whole in a field known as organismic biology; that area of research remains an important component of the biological sciences. Population biology deals with groups or populations of organisms that inhabit a given area or region. Included at that level are studies of the roles that specific kinds of plants and animals play in the complex and self-perpetuating interrelationships that exist between the living and the nonliving world, as well as studies of the built-in controls that maintain those relationships naturally. Those broadly based levels—molecules, cells, whole organisms, and populations—may be further subdivided for study, giving rise to specializations such as morphology, taxonomy, biophysics, biochemistry, genetics, epigenetics, and ecology. A field of biology may be especially concerned with the investigation of one kind of living thing—for example, the study of birds in ornithology, the study of fishes in ichthyology, or the study of microorganisms in microbiology.

Biological Principles

Unity

All living organisms, regardless of their uniqueness, have certain biological, chemical, and physical characteristics in common. All, for example, are composed of basic units known as cells and of the same chemical substances, which, when analyzed, exhibit noteworthy similarities, even in such disparate organisms as bacteria and humans. Furthermore, since the action of any organism is determined by the manner in which its cells interact and since all cells interact in much the same way, the basic functioning of all organisms is also similar.

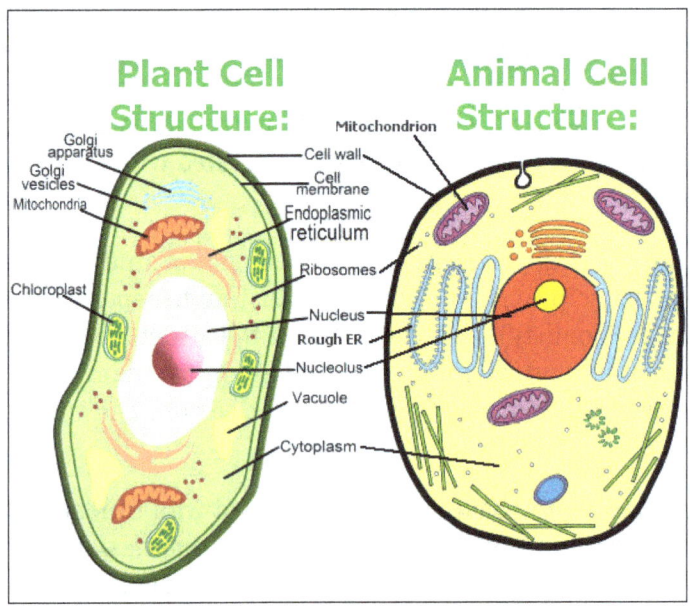

There is not only unity of basic living substance and functioning but also unity of origin of all living things. According to a theory proposed in 1855 by German pathologist Rudolf Virchow, "all living cells arise from pre-existing living cells." That theory appears to be true for all living things at the present time under existing environmental conditions. If, however, life originated on Earth more than once in the past, the fact that all organisms have a sameness of basic structure, composition, and function would seem to indicate that only one original type succeeded.

A common origin of life would explain why in humans or bacteria—and in all forms of life in between—the same chemical substance, deoxyribonucleic acid (DNA), in the form of genes accounts for the ability of all living matter to replicate itself exactly and to transmit genetic information from parent to offspring. Furthermore, the mechanisms for that transmittal follow a pattern that is the same in all organisms.

Whenever a change in a gene (a mutation) occurs, there is a change of some kind in the organism that contains the gene. It is this universal phenomenon that gives rise to the differences (variations) in populations of organisms from which nature selects for survival those that are best able to cope with changing conditions in the environment.

Behaviour and Interrelationships

The study of the relationships of living things to each other and to their environment is known as ecology. Because these interrelationships are so important to the welfare of Earth and because they can be seriously disrupted by human activities, ecology has become an important branch of biology.

Continuity

Whether an organism is a human or a bacterium, its ability to reproduce is one of the most important characteristics of life. Because life comes only from preexisting life, it is only through reproduction that successive generations can carry on the properties of a species.

Study of Structure

Living things are defined in terms of the activities or functions that are missing in nonliving things. The life processes of every organism are carried out by specific materials assembled in definite structures. Thus, a living thing can be defined as a system, or structure, that reproduces, changes with its environment over a period of time, and maintains its individuality by constant and continuous metabolism.

Cells and their Constituents

Biologists once depended on the light microscope to study the morphology of cells found in higher plants and animals. The functioning of cells in unicellular and in multicellular organisms was then postulated from observation of the structure; the discovery of the chloroplastids in the cell, for example, led to the investigation of the process of photosynthesis. With the invention of the electron microscope, the fine organization of the plastids could be used for further quantitative studies of the different parts of that process.

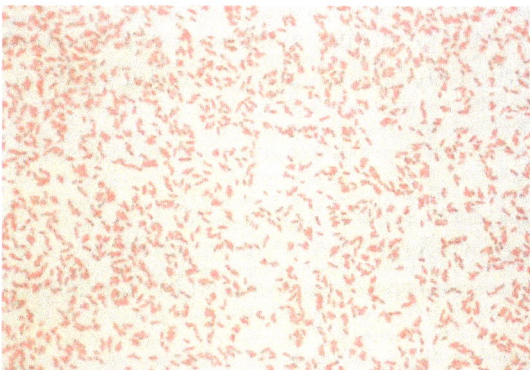

Yersinia enterocolitica: Photomicrograph of Gram stain of Yersinia enterocolitica, the causative agent of yersiniosis.

Qualitative and quantitative analyses in biology make use of a variety of techniques and approaches to identify and estimate levels of nucleic acids, proteins, carbohydrates, and other chemical constituents of cells and tissues. Many such techniques make use of antibodies or probes that bind to specific molecules within cells and that are tagged with a chemical, commonly a fluorescent dye, a radioactive isotope, or a biological stain, thereby enabling or enhancing microscopic visualization or detection of the molecules of interest.

Chemical labels are powerful means by which biologists can identify, locate, or trace substances in living matter. Some examples of widely used assays that incorporate labels include the Gram stain, which is used for the identification and characterization of bacteria; fluorescence in situ hybridization, which is used for the detection of specific genetic sequences in chromosomes; and luciferase assays, which measure bioluminescence produced from luciferin-luciferase reactions, allowing for the quantification of a wide array of molecules.

Tissues and Organs

Early biologists viewed their work as a study of the organism. The organism, then considered the fundamental unit of life, is still the prime concern of some modern biologists, and understanding how organisms maintain their internal environment remains an important part of biological research. To better understand the physiology of organisms, researchers study the tissues and organs of which organisms are composed. Key to that work is the ability to maintain and grow cells in vitro ("in glass"), otherwise known as tissue culture.

Some of the first attempts at tissue culture were made in the late 19th century. In 1885, German zoologist Wilhelm Roux maintained tissue from a chick embryo in a salt solution. The first major breakthrough in tissue culture, however, came in 1907 with the growth of frog nerve cell processes by American zoologist Ross G. Harrison. Several years later, French researchers Alexis Carrel and Montrose Burrows had refined Harrison's methods and introduced the term tissue culture. Using stringent laboratory techniques, workers have been able to keep cells and tissues alive under culture conditions for long periods of time. Techniques for keeping organs alive in preparation for transplants stem from such experiments.

Advances in tissue culture have enabled countless discoveries in biology. For example, many experiments have been directed toward achieving a deeper understanding of biological differentiation, particularly of the factors that control differentiation. Crucial to those studies was the development in the late 20th century of tissue culture methods that allowed for the growth of mammalian embryonic stem cells—and ultimately human embryonic stem cells—on culture plates.

Properties of Life

All groups of living organisms share several key characteristics or functions: order, sensitivity or response to stimuli, reproduction, adaptation, growth and development, regulation, homeostasis, and energy processing. When viewed together, these eight characteristics serve to define life.

Order

Organisms are highly organized structures that consist of one or more cells. Even very simple, single-celled organisms are remarkably complex. Inside each cell, atoms make up molecules. These in turn make up cell components or organelles. Multicellular organisms, which may consist of millions of individual cells, have an advantage over single-celled organisms in that their cells can be specialized to perform specific functions, and even sacrificed in certain situations for the good of the organism as a whole.

Figure: A toad represents a highly organized structure consisting of cells, tissues, organs, and organ systems.

Sensitivity or Response to Stimuli

Figure: The leaves of this sensitive plant (Mimosa pudica) will instantly droop and fold when touched. After a few minutes, the plant returns to its normal state.

Organisms respond to diverse stimuli. For example, plants can bend toward a source of light or respond to touch. Even tiny bacteria can move toward or away from chemicals (a process called chemotaxis) or light (phototaxis). Movement toward a stimulus is considered a positive response, while movement away from a stimulus is considered a negative response.

Reproduction

Single-celled organisms reproduce by first duplicating their DNA, which is the genetic material, and then dividing it equally as the cell prepares to divide to form two new cells. Many multicellular organisms (those made up of more than one cell) produce specialized reproductive cells that will form new individuals. When reproduction occurs, DNA containing genes is passed along to an organism's offspring. These genes are the reason that the offspring will belong to the same species and will have characteristics similar to the parent, such as fur color and blood type.

Adaptation

All living organisms exhibit a "fit" to their environment. Biologists refer to this fit as adaptation and it is a consequence of evolution by natural selection, which operates in every lineage of reproducing organisms. Examples of adaptations are diverse and unique, from heat-resistant Archaea that live in boiling hot springs to the tongue length of a nectar-feeding moth that matches the size of the flower from which it feeds. All adaptations enhance the reproductive potential of the individual exhibiting them, including their ability to survive to reproduce. Adaptations are not constant. As an environment changes, natural selection causes the characteristics of the individuals in a population to track those changes.

Growth and Development

Figure: Although no two look alike, these kittens have inherited genes from both parents and share many of the same characteristics.

Organisms grow and develop according to specific instructions coded for by their genes. These genes provide instructions that will direct cellular growth and development, ensuring that a species' young will grow up to exhibit many of the same characteristics as its parents.

Regulation

Even the smallest organisms are complex and require multiple regulatory mechanisms to coordinate internal functions, such as the transport of nutrients, response to stimuli, and coping with environmental stresses. For example, organ systems such as the digestive or circulatory systems perform specific functions like carrying oxygen throughout the body, removing wastes, delivering nutrients to every cell, and cooling the body.

Homeostasis

Figure: Polar bears and other mammals living in ice-covered regions maintain their body temperature by generating heat and reducing heat loss through thick fur and a dense layer of fat under their skin.

To function properly, cells require appropriate conditions such as proper temperature, pH, and concentrations of diverse chemicals. These conditions may, however, change from one moment to the next. Organisms are able to maintain internal conditions within a narrow range almost constantly, despite environmental changes, through a process called homeostasis or "steady state"—the ability of an organism to maintain constant internal conditions. For example, many organisms regulate their body temperature in a process known as thermoregulation. Organisms that live in cold climates, such as the polar bear have body structures that help them withstand low temperatures and conserve body heat. In hot climates, organisms have methods (such as perspiration in humans or panting in dogs) that help them to shed excess body heat.

Energy Processing

Figure: A lot of energy is required for a California condor to fly. Chemical energy derived from food is used to power flight. California condors are an endangered species; scientists have strived to place a wing tag on each bird to help them identify and locate each individual bird.

All organisms (such as the California condor) use a source of energy for their metabolic activities. Some organisms capture energy from the Sun and convert it into chemical energy in food; others use chemical energy from molecules they take in.

Levels of Organization of Living Things

Living things are highly organized and structured, following a hierarchy on a scale from small to large. The atom is the smallest and most fundamental unit of matter. It consists of a nucleus surrounded by electrons. Atoms form molecules. A molecule is a chemical structure consisting of at least two atoms held together by a chemical bond. Many molecules that are biologically important are macromolecules, large molecules that are typically formed by combining smaller units called monomers. An example of a macromolecule is deoxyribonucleic acid (DNA), which contains the instructions for the functioning of the organism that contains it.

Some cells contain aggregates of macromolecules surrounded by membranes; these are called organelles. Organelles are small structures that exist within cells and perform specialized functions. All living things are made of cells; the cell itself is the smallest fundamental unit of structure and function in living organisms. (This requirement is why viruses are not considered living: they are not made of cells. To make new viruses, they have to invade and hijack a living cell; only then can they obtain the materials they need to reproduce.) Some organisms consist of a single cell and others are multicellular. Cells are classified as prokaryotic or eukaryotic. Prokaryotes are single-celled organisms that lack organelles surrounded by a membrane and do not have nuclei surrounded by nuclear membranes; in contrast, the cells of eukaryotes do have membrane-bound organelles and nuclei.

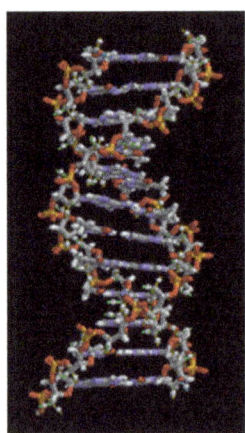

Figure: A molecule, like this large DNA molecule, is composed of atoms.

In most multicellular organisms, cells combine to make tissues, which are groups of similar cells carrying out the same function. Organs are collections of tissues grouped together based on a common function. Organs are present not only in animals but also in plants. An organ system is a higher level of organization that consists of functionally related organs. For example vertebrate animals have many organ systems, such as the circulatory system that transports blood throughout the body and to and from the lungs; it includes organs such as the heart and blood vessels. Organisms are individual living entities. For example, each tree in a forest is an organism. Single-celled prokaryotes and single-celled eukaryotes are also considered organisms and are typically referred to as microorganisms.

- Atom: A basic unit of matter that consists of dense central nucleus surrounded by a cloud of negatively charged electrons.

- Molecule: A phospholipid, composed of many atoms.

- Organelles: Structures that perform functions within a cell. Highlighted in blue are a Golgi apparatus and a nucleus.

- Cells: Human blood cells.

- Tissue: Human skin tissue.

- Organs and organ systems: Organs such as the stomach and intestine make up part of the human digestive system.

- Organisms, populations, and communities: In a park, each person is an organism. Together, all the people make up a population. All the plant and animal species in the park comprise a community.

- Ecosystem: The ecosystem of Central Park in New York includes living organisms and the environment in which they live.

- The biosphere: Encompasses all the ecosystem on Earth.

All the individuals of a species living within a specific area are collectively called a population. For example, a forest may include many white pine trees. All of these pine trees represent the population of white pine trees in this forest. Different populations may live in the same specific area. For example, the forest with the pine trees includes populations of flowering plants and also insects and microbial populations. A community is the set of populations inhabiting a particular area. For instance, all of the trees, flowers, insects, and other populations in a forest form the forest's community. The forest itself is an ecosystem. An ecosystem consists of all the living things in a particular area together with the abiotic, or non-living, parts of that environment such as nitrogen in the soil or rainwater. At the highest level of organization the biosphere is the collection of all ecosystems, and it represents the zones of life on Earth. It includes land, water, and portions of the atmosphere.

The Diversity of Life

Despite the basic biological, chemical, and physical similarities found in all living things, a diversity of life exists not only among and between species but also within every natural population. The phenomenon of diversity has had a long history of study because so many of the variations that exist in nature are visible to the eye. The fact that organisms changed during prehistoric times and that new variations are constantly evolving can be verified by paleontological records as well as by breeding experiments in the laboratory. Long after Darwin assumed that variations existed, biologists discovered that they are caused by a change in the genetic material (DNA). That change can be a slight alteration in the sequence of the constituents of DNA (nucleotides), a larger change such as a structural alteration of a chromosome, or a complete change in the number of chromosomes. In any case, a change in the genetic material in the reproductive cells manifests itself as some kind of structural or chemical change in the offspring. The consequence of such a mutation depends upon the interaction of the mutant offspring with its environment.

It has been suggested that sexual reproduction became the dominant type of reproduction among organisms because of its inherent advantage of variability, which is the mechanism that enables a species to adjust to changing conditions. New variations are potentially present in genetic differences, but how preponderant a variation becomes in a gene pool depends upon the number of offspring the mutants or variants produce (differential reproduction). It is possible for a genetic novelty (new variation) to spread in time to all members of a population, especially if the novelty enhances the population's chances for survival in the environment in which it exists. Thus, when a species is introduced into a new habitat, it either adapts to the change by natural selection or by some other evolutionary mechanism or eventually dies off. Because each new habitat means new adaptations, habitat changes have been responsible for the millions of different kinds of species and for the heterogeneity within each species.

The total number of extant animal and plant species is estimated at between roughly 5 million and 10 million; about 1.5 million of those species have been described by scientists. The use of classification as a means of producing some kind of order out of the staggering number of different types of organisms appeared as early as the book of Genesis—with references to cattle, beasts, fowl, creeping things, trees, and so on. The first scientific attempt at classification, however, is attributed to the Greek philosopher Aristotle, who tried to establish a system that would indicate the relationship of all things to each other. He arranged everything along a scale, or "ladder of nature," with nonliving things at the bottom; plants were placed below animals, and humankind was at the top. Other schemes that have been used for grouping species include large anatomical similarities, such as wings or fins, which indicate a natural relationship, and also similarities in reproductive structures.

Taxonomy has been based on two major assumptions: one is that similar body construction can be used as a criterion for a classification grouping; the other is that, in addition to structural similarities, evolutionary and molecular relationships between organisms can be used as a means for determining classification.

The science of biology is very broad in scope because there is a tremendous diversity of life on Earth. The source of this diversity is evolution, the process of gradual change during which new species arise from older species. Evolutionary biologists study the evolution of living things in everything from the microscopic world to ecosystems.

In the 18th century, a scientist named Carl Linnaeus first proposed organizing the known species of organisms into a hierarchical taxonomy. In this system, species that are most similar to each other are put together within a grouping known as a genus. Furthermore, similar genera (the plural of genus) are put together within a family. This grouping continues until all organisms are collected together into groups at the highest level. The current taxonomic system now has eight levels in its hierarchy, from lowest to highest, they are: species, genus, family, order, class, phylum, kingdom, domain. Thus species are grouped within genera, genera are grouped within families, families are grouped within orders, and so on.

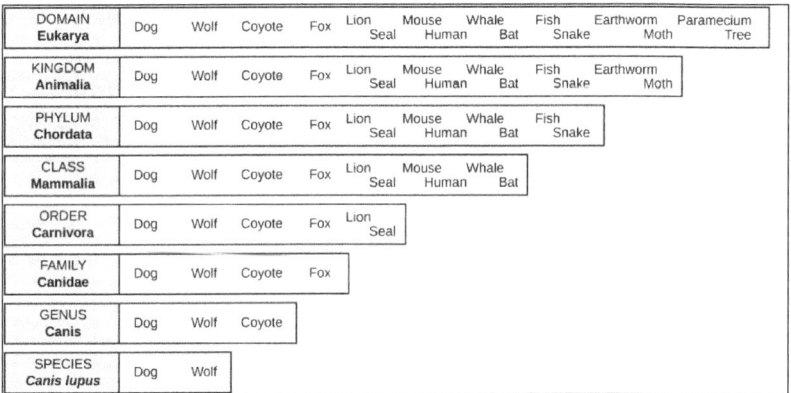

Figure: This diagram shows the levels of taxonomic hierarchy for a dog, from the broadest category—domain—to the most specific—species.

The highest level, domain, is a relatively new addition to the system since the 1990s. Scientists now recognize three domains of life, the Eukarya, the Archaea, and the Bacteria. The domain Eukarya contains organisms that have cells with nuclei. It includes the kingdoms of fungi, plants, animals, and several kingdoms of protists. The Archaea, are single-celled organisms without nuclei and include many extremophiles that live in harsh environments like hot springs. The Bacteria are another quite different group of single-celled organisms without nuclei. Both the Archaea and the Bacteria are prokaryotes, an informal name for cells without nuclei. The recognition in the 1990s that certain "bacteria," now known as the Archaea, were as different genetically and biochemically from other bacterial cells as they were from eukaryotes, motivated the recommendation to divide life into three domains. This dramatic change in our knowledge of the tree of life demonstrates that classifications are not permanent and will change when new information becomes available.

In addition to the hierarchical taxonomic system, Linnaeus was the first to name organisms using two unique names, now called the binomial naming system. Before Linnaeus, the use of common names to refer to organisms caused confusion because there were regional differences in these common names. Binomial names consist of the genus name (which is capitalized) and the species name (all lower-case). Both names are set in italics when they are printed. Every species is given a unique binomial which is recognized the world over, so that a scientist in any location can know which organism is being referred to. For example, the North American blue jay is known uniquely as Cyanocitta cristata. Our own species is Homo sapiens.

Figure: These images represent different domains. The scanning electron micrograph shows (a) bacterial cells belong to the domain Bacteria, while the (b) extremophiles, seen all together as colored mats in this hot spring, belong to domain Archaea. Both the (c) sunflower and (d) lion are part of domain Eukarya.

Evolution in Action

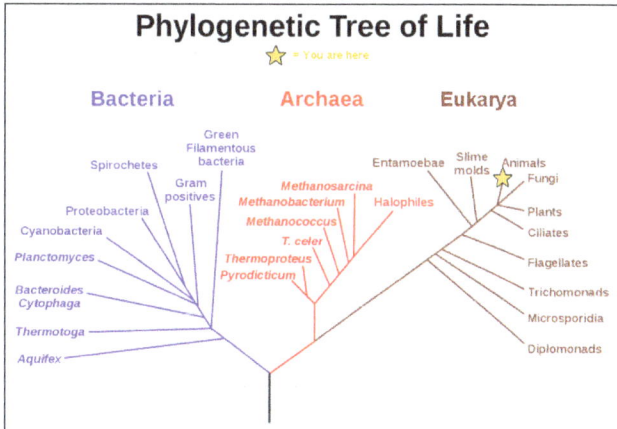

Figure: This phylogenetic tree was constructed by microbiologist Carl Woese using genetic relationships. The tree shows the separation of living organisms into three domains: Bacteria, Archaea, and Eukarya. Bacteria and Archaea are organisms without a nucleus or other organelles surrounded by a membrane and, therefore, are prokaryotes.

Carl Woese and the Phylogenetic Tree

The evolutionary relationships of various life forms on Earth can be summarized in a phylogenetic tree. A phylogenetic tree is a diagram showing the evolutionary relationships among biological species based on similarities and differences in genetic or physical traits or both. A phylogenetic tree is composed of branch points, or nodes, and branches. The internal nodes represent ancestors and are points in evolution when, based on scientific evidence, an ancestor is thought to have diverged to form two new species. The length of each branch can be considered as estimates of relative time.

In the past, biologists grouped living organisms into five kingdoms: animals, plants, fungi, protists, and bacteria. The pioneering work of American microbiologist Carl Woese in the early 1970s has shown, however, that life on Earth has evolved along three lineages, now called domains—Bacteria, Archaea, and Eukarya. Woese proposed the domain as a new taxonomic level and Archaea as a new domain, to reflect the new phylogenetic tree. Many organisms belonging to the Archaea domain live under extreme conditions and are called extremophiles. To construct his tree, Woese used genetic relationships rather than similarities based on morphology (shape). Various genes were used in phylogenetic studies. Woese's tree was constructed from comparative sequencing of the genes that are universally distributed, found in some slightly altered form in every organism, conserved (meaning that these genes have remained only slightly changed throughout evolution), and of an appropriate length.

Branches of Biology

The scope of biology is broad and therefore contains many branches and sub disciplines. Biological science is classified into various branches, depending upon the organisms to be studied, and is a vast field.

Zoology

This is a branch of biology that studies animals. Zoology is divided into Applied Zoology, the study of production and non production animals, Systematic Zoology, dealing with evolution and taxonomy or science of naming living things and Organismal Zoology, the study of animals in our biosphere. Applied Zoology is further divided into, Aquaculture, which involves production and maintenance of freshwater and seawater animals and plants, Piggery, which includes study of everything related to pigs, Applied Entomology, which includes manipulation of insects for the benefit of humans, Vermiculture, which is breeding of the worms which burrow soil, for production of natural fertilizers, Poultry Science, the study of domestic birds such as geese, turkey and chicken, Parasitology, dealing with the study of parasites, Radiation Biology, which uses gamma rays, X-rays, electrons and protons for well-being of humans, Biotechnology, which applies engineering principles for the material processing by biological factors, Applied Embryology, which embraces test tube culture (embryo culture) for increasing productivity from cattle, Tissue Culture, involving the culture of plant tissues and cells in an artificial environment, Dairy Science, which deals with milk or milk related products, Pesticide Technology, which is the study of pesticides and their uses, Nematology which deals with study of roundworms of organisms and their control, Ornithology, which is the study of birds, Herpetology, study of reptiles, Ichthyology, which is the study of fish and Mammology, which includes the study of mammals.

Entomology

One of the sub branches is entomology, which is exclusively based on insects. It concentrates on studying the taxonomy, features, adaptations, roles and behavior of insects.

Ethology

Ethology comes under zoology and deals with behavioral adaptations of animals, specially in their natural or original dwelling places.

Anatomy

Applicable to plant anatomy and animal anatomy, it involves studying the detailed structure, internal organs and the respective functions of an organism.

Physiology

Physiology is defined as the study of various functions and processes of living organisms. Physiology is further divided into Evolutionary Physiology, which is the study of physiological evolution, Cell Physiology - the study of cell mechanism and interaction, Developmental Physiology, which involves the study of physiological processes in relation to embryonic evolution, Environmental Physiology, which deals with the study of response of plants to agents such as temperature, radiation and fire and Comparative Physiology, roughly explained as the study of animals except humans.

Genetics

This is considered to be an interesting field of study and is a branch of biology. Genetics is the study of genes. This term is derived from the Greek word "genetikos" meaning "origin". This branch of biology studies about the hereditary aspects of all living organisms. The study of inheritance of traits from the parent had begun in the mid-nineteenth century and was pioneered by a renowned biologist Gregor Mendel. The modern science of genetics is based upon the foundations laid down by this biologist.

Botany

The study of plant life or phytology is known as botany. One of the most prominent among the different branches of biology, botany is a vast subject and studies the life and development of fungi, algae and plants. Botany also probes into the structure, growth, diseases, chemical and physical properties, metabolism and evolution of the plant species. Botany implies the importance of study of plant life on earth because they generate food, fibers, medicines, fuel and oxygen.

Evolution Biology

As we all know, highly developed organism have evolved from simpler forms. There is a specific branch of biology, called evolution biology that focuses on the evolution of species.

Developmental Biology

As the name signifies, development biology helps a student in learning the various phases of growth and development of a living creature.

Ecology

Ecology is a branch of biology that studies the interaction of various organisms with one another, and their chemical and physical environment. This branch of biology studies environmental problems such as pollution and how it affects the eco-cycle. The term ecology is derived from the Greek term "oikos" meaning "household" and "logos" meaning "study". A German biologist, Ernst Haeckel, coined the term ecology in 1866.

Cryobiology

This deals with the effects of extremely low temperature in living cells and organisms as a whole.

Biochemistry

This branch of biology studies the chemical processes in all living organisms. Biochemistry is a branch of science that studies the functions of the cellular components such as nucleic acids, lipids, proteins and various other bio-molecules.

Cytology and Molecular Biology

In-depth study about the cell along with its structure, function, parts and abnormalities are all studied under cell biology or cytology. Likewise, study of organisms at the molecular level is called molecular biology.

Marine Biology

Marine biology studies the ecosystem of the oceans, marine animals and plants. There is a vast portion of ocean life that is still unexplored. You can rightly say that marine biology is a branch of oceanography, which is, again, a branch of biology.

Bioinformatics

Bioinformatics basically relates to genomic studies with the application of data processing, computational knowledge and statistical applications.

Mycology

According to modern-day taxonomy, fungi (singular fungus) is neither a plant nor an animal. It belongs to a different living group and is studied under the subject, mycology.

Biophysics

Biophysics involves the study of relation between organisms or living cells and electrical or mechanical energy. Biophysics is further divided into the following sub-branches: Molecular Biophysics, which defines biological functions in relation to dynamic behavior and molecular structure of various living systems such as viruses, Bio mechanics is the study of forces applied by muscles and gravity on the skeleton, Bio electricity - the study of electric currents flowing through muscles and nerves and static voltage of biological cells, Cellular Biophysics, which incorporates study of membrane function and structure, and cellular excitation and Quantum Biophysics, which includes the study of behavior of living matter at molecular and sub molecular level.

Aquatic Biology

It involves study of life in water, like study of various species of animals, plants and micro-organisms. It incorporates the study of both freshwater and sea water organisms. Sometimes, aquatic biology is also referred to as limnology.

References

- Biology, science: britannica.com, Retrieved 4 March, 2019
- The-science-of-biology-and-the-study-of-life: brewminate.com, Retrieved 12 January, 2019
- Evolution, biology, science: britannica.com, Retrieved 8 April, 2019
- The-science-of-biology-and-the-study-of-life: brewminate.com, Retrieved 11 July, 2019
- Different-branches-of-biology: biologywise.com, Retrieved 20 February, 2019

Chapter 2
Cell: The Fundamental Unit of Life

The cell is the fundamental and smallest unit of life of all known living organisms. There are two types of cells- eukaryotic and prokaryotic cells. Cellular respiration and cellular reproduction are the two main functions of cells. The chapter closely examines these key concepts related to cells to provide an extensive understanding of the subject.

Cell

In Biology, Cell is the basic membrane-bound unit that contains the fundamental molecules of life and of which all living things are composed. A single cell is often a complete organism in itself, such as a bacterium or yeast. Other cells acquire specialized functions as they mature. These cells cooperate with other specialized cells and become the building blocks of large multicellular organisms, such as humans and other animals. Although cells are much larger than atoms, they are still very small. The smallest known cells are a group of tiny bacteria called mycoplasmas; some of these single-celled organisms are spheres as small as 0.2 µm in diameter (1µm = about 0.000039 inch), with a total mass of 10–14 gram—equal to that of 8,000,000,000 hydrogen atoms. Cells of humans typically have a mass 400,000 times larger than the mass of a single mycoplasma bacterium, but even human cells are only about 20 µm across. It would require a sheet of about 10,000 human cells to cover the head of a pin, and each human organism is composed of more than 75,000,000,000,000 cells.

As an individual unit, the cell is capable of metabolizing its own nutrients, synthesizing many types of molecules, providing its own energy, and replicating itself in order to produce succeeding generations. It can be viewed as an enclosed vessel, within which innumerable chemical reactions take place simultaneously. These reactions are under very precise control so that they contribute to the life and procreation of the cell. In a multicellular organism, cells become specialized to perform different functions through the process of differentiation. In order to do this, each cell keeps in constant communication with its neighbours. As it receives nutrients from and expels wastes into its surroundings, it adheres to and cooperates with other cells. Cooperative assemblies of similar cells form tissues, and a cooperation between tissues in turn forms organs, which carry out the functions necessary to sustain the life of an organism.

The Nature and Function of Cells

A cell is enclosed by a plasma membrane, which forms a selective barrier that allows nutrients to enter and waste products to leave. The interior of the cell is organized into many specialized compartments, or organelles, each surrounded by a separate membrane. One major organelle, the nucleus, contains the genetic information necessary for cell growth and reproduction. Each cell contains only one nucleus, whereas other types of organelles are present in multiple copies in the cellular contents, or cytoplasm. Organelles include mitochondria, which are responsible for the energy transactions necessary for cell survival; lysosomes, which digest unwanted materials within the cell; and the endoplasmic reticulum and the Golgi apparatus, which play important roles in the internal organization of the cell by synthesizing selected molecules and then processing, sorting, and directing them to their proper locations. In addition, plant cells contain chloroplasts, which are responsible for photosynthesis, whereby the energy of sunlight is used to convert molecules of carbon dioxide (CO_2) and water (H_2O) into carbohydrates. Between all these organelles is the space in the cytoplasm called the cytosol. The cytosol contains an organized framework of fibrous molecules that constitute the cytoskeleton, which gives a cell its shape, enables organelles to move within the cell, and provides a mechanism by which the cell itself can move. The cytosol also contains more than 10,000 different kinds of molecules that are involved in cellular biosynthesis, the process of making large biological molecules from small ones.

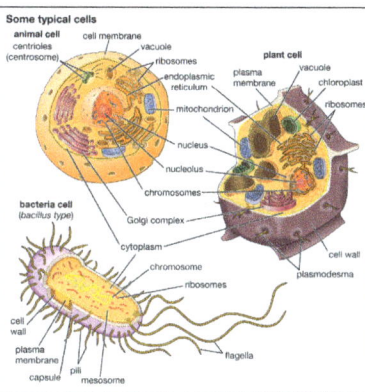

Animal cells and plant cells contain membrane-bound organelles, including a distinct nucleus. In contrast, bacterial cells do not contain organelles.

Specialized organelles are a characteristic of cells of organisms known as eukaryotes. In contrast, cells of organisms known as prokaryotes do not contain organelles and are generally smaller than eukaryotic cells. However, all cells share strong similarities in biochemical function.

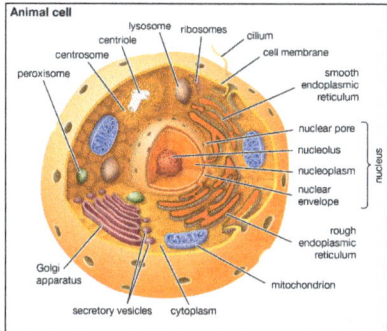

Cutaway drawing of a eukaryotic cell.

Molecules of Cells

Cells contain a special collection of molecules that are enclosed by a membrane. These molecules give cells the ability to grow and reproduce. The overall process of cellular reproduction occurs in two steps: cell growth and cell division. During cell growth, the cell ingests certain molecules from its surroundings by selectively carrying them through its cell membrane. Once inside the cell, these molecules are subjected to the action of highly specialized, large, elaborately folded molecules called enzymes. Enzymes act as catalysts by binding to ingested molecules and regulating the rate at which they are chemically altered. These chemical alterations make the molecules more useful to the cell. Unlike the ingested molecules, catalysts are not chemically altered themselves during the reaction, allowing one catalyst to regulate a specific chemical reaction in many molecules.

Biological catalysts create chains of reactions. In other words, a molecule chemically transformed by one catalyst serves as the starting material, or substrate, of a second catalyst and so on. In this way, catalysts use the small molecules brought into the cell from the outside environment to create increasingly complex reaction products. These products are used for cell growth and the replication of genetic material. Once the genetic material has been copied and there are sufficient molecules to support cell division, the cell divides to create two daughter cells. Through many such cycles of cell growth and division, each parent cell can give rise to millions of daughter cells, in the process converting large amounts of inanimate matter into biologically active molecules.

The Structure of Biological Molecules

Cells are largely composed of compounds that contain carbon. The study of how carbon atoms interact with other atoms in molecular compounds forms the basis of the field of organic chemistry and plays a large role in understanding the basic functions of cells. Because carbon atoms can form stable bonds with four other atoms, they are uniquely suited for the construction of complex molecules. These complex molecules are typically made up of chains and rings that contain hydrogen, oxygen, and nitrogen atoms, as well as carbon atoms. These molecules may consist of anywhere from 10 to millions of atoms linked together in specific arrays. Most, but not all, of the carbon-containing molecules in cells are built up from members of one of four different families of small organic molecules: sugars, amino acids, nucleotides, and fatty acids. Each of these families contains a group of molecules that resemble one another in both structure and function. In addition to other important functions, these molecules are used to build large macromolecules. For example, the sugars can be linked to form polysaccharides such as starch and glycogen, the amino acids can be linked to form proteins, the nucleotides can be linked to form the DNA (deoxyribonucleic acid) and RNA (ribonucleic acid) of chromosomes, and the fatty acids can be linked to form the lipids of all cell membranes.

component	percent of total cell weight
water	70
inorganic ions (sodium, potassium, magnesium, calcium, chloride, etc.)	1
miscellaneous small metabolites	3
proteins	18
RNA	1.1
DNA	0.25
phospholipids and other lipids	5

component	percent of total cell weight
water	70
polysaccharides	2
Approximate chemical composition of a typical mammalian cell	

Aside from water, which forms 70 percent of a cell's mass, a cell is composed mostly of macromolecules. By far the largest portion of macromolecules are the proteins. An average-sized protein macromolecule contains a string of about 400 amino acid molecules. Each amino acid has a different side chain of atoms that interact with the atoms of side chains of other amino acids. These interactions are very specific and cause the entire protein molecule to fold into a compact globular form. In theory, nearly an infinite variety of proteins can be formed, each with a different sequence of amino acids. However, nearly all these proteins would fail to fold in the unique ways required to form efficient functional surfaces and would therefore be useless to the cell. The proteins present in cells of modern animals and humans are products of a long evolutionary history, during which the ancestor proteins were naturally selected for their ability to fold into specific three-dimensional forms with unique functional surfaces useful for cell survival.

Most of the catalytic macromolecules in cells are enzymes. The majority of enzymes are proteins. Key to the catalytic property of an enzyme is its tendency to undergo a change in its shape when it binds to its substrate, thus bringing together reactive groups on substrate molecules. Some enzymes are macromolecules of RNA, called ribozymes. Ribozymes consist of linear chains of nucleotides that fold in specific ways to form unique surfaces, similar to the ways in which proteins fold. As with proteins, the specific sequence of nucleotide subunits in an RNA chain gives each macromolecule a unique character. RNA molecules are much less frequently used as catalysts in cells than are protein molecules, presumably because proteins, with the greater variety of amino acid side chains, are more diverse and capable of complex shape changes. However, RNA molecules are thought to have preceded protein molecules during evolution and to have catalyzed most of the chemical reactions required before cells could evolve.

The Genetic Information of Cells

Cells can thus be seen as a self-replicating network of catalytic macromolecules engaged in a carefully balanced series of energy conversions that drive biosynthesis and cell movement. But energy alone is not enough to make self-reproduction possible; the cell must contain detailed instructions that dictate exactly how that energy is to be used. These instructions are analogous to the blueprints that a builder uses to construct a house; in the case of cells, however, the blueprints themselves must be duplicated along with the cell before it divides, so that each daughter cell can retain the instructions that it needs for its own replication. These instructions constitute the cell's heredity.

DNA: The Genetic Material

During the early 19th century, it became widely accepted that all living organisms are composed of cells arising only from the growth and division of other cells. The improvement of the microscope then led to an era during which many biologists made intensive observations of the microscopic structure of cells. By 1885 a substantial amount of indirect evidence indicated that chromosomes—dark-staining threads in the cell nucleus—carried the information for cell heredity. It was later shown that chromosomes are about half DNA and half protein by weight.

The revolutionary discovery suggesting that DNA molecules could provide the information for their own replication came in 1953, when American geneticist and biophysicist James Watson and British biophysicist Francis Crick proposed a model for the structure of the double-stranded DNA molecule (called the DNA double helix). In this model, each strand serves as a template in the synthesis of a complementary strand. Subsequent research confirmed the Watson and Crick model of DNA replication and showed that DNA carries the genetic information for reproduction of the entire cell.

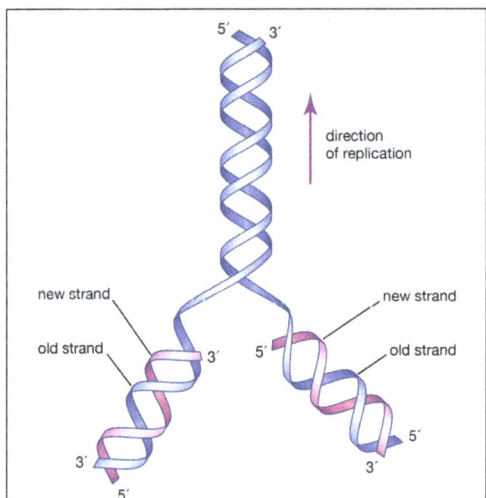

The initial proposal of the structure of DNA by James Watson and Francis Crick was accompanied by a suggestion on the means of replication.

All of the genetic information in a cell was initially thought to be confined to the DNA in the chromosomes of the cell nucleus. Later discoveries identified small amounts of additional genetic information present in the DNA of much smaller chromosomes located in two types of organelles in the cytoplasm. These organelles are the mitochondria in animal cells and the mitochondria and chloroplasts in plant cells. The special chromosomes carry the information coding for a few of the many proteins and RNA molecules needed by the organelles. They also hint at the evolutionary origin of these organelles, which are thought to have originated as free-living bacteria that were taken up by other organisms in the process of symbiosis.

RNA: Replicated from DNA

It is possible for RNA to replicate itself by mechanisms related to those used by DNA, even though it has a single-stranded instead of a double-stranded structure. In early cells RNA is thought to have replicated itself in this way. However, all of the RNA in present-day cells is synthesized by special enzymes that construct a single-stranded RNA chain by using one strand of the DNA helix as a template. Although RNA molecules are synthesized in the cell nucleus, where the DNA is located, most of them are transported to the cytoplasm before they carry out their functions.

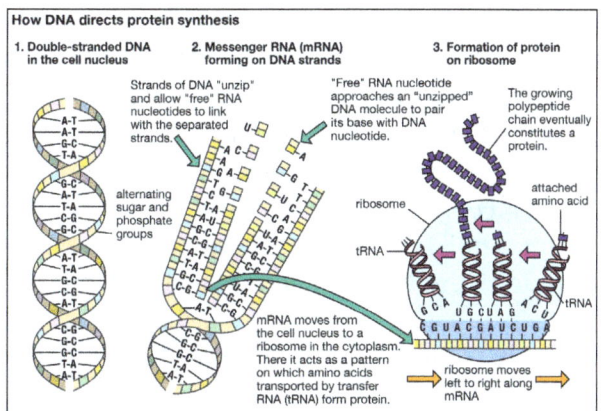

Molecular genetics emerged from the realization that DNA and RNA constitute the genetic material of all living organisms. (1) DNA, located in the cell nucleus, is made up of nucleotides that contain the bases adenine (A), thymine (T), guanine (G), and cytosine (C). (2) RNA, which contains uracil (U) instead of thymine, transports the genetic code to protein-synthesizing sites in the cell. (3) Messenger RNA (mRNA) then carries the genetic information to ribosomes in the cell cytoplasm that translate the genetic information into molecules of protein.

The RNA molecules in cells have two main roles. Some, the ribozymes, fold up in ways that allow them to serve as catalysts for specific chemical reactions. Others serve as "messenger RNA," which provides templates specifying the synthesis of proteins. Ribosomes, tiny protein-synthesizing machines located in the cytoplasm, "read" the messenger RNA molecules and "translate" them into proteins by using the genetic code. In this translation, the sequence of nucleotides in the messenger RNA chain is decoded three nucleotides at a time, and each nucleotide triplet (called a codon) specifies a particular amino acid. Thus, a nucleotide sequence in the DNA specifies a protein provided that a messenger RNA molecule is produced from that DNA sequence. Each region of the DNA sequence specifying a protein in this way is called a gene.

By the above mechanisms, DNA molecules catalyze not only their own duplication but also dictate the structures of all protein molecules. A single human cell contains about 10,000 different proteins produced by the expression of 10,000 different genes. Actually, a set of human chromosomes is thought to contain DNA with enough information to express between 30,000 and 100,000 proteins, but most of these proteins seem to be made only in specialized types of cells and are therefore not present throughout the body.

The Organization of Cells

Intracellular Communication

A cell with its many different DNA, RNA, and protein molecules is quite different from a test tube containing the same components. When a cell is dissolved in a test tube, thousands of different types of molecules randomly mix together. In the living cell, however, these components are kept in specific places, reflecting the high degree of organization essential for the growth and division of the cell. Maintaining this internal organization requires a continuous input of energy, because spontaneous chemical reactions always create disorganization. Thus, much of the energy released by ATP hydrolysis fuels processes that organize macromolecules inside the cell.

When a eukaryotic cell is examined at high magnification in an electron microscope, it becomes apparent that specific membrane-bound organelles divide the interior into a variety of subcompartments. Although not detectable in the electron microscope, it is clear from biochemical assays that each organelle contains a different set of macromolecules. This biochemical segregation reflects the functional specialization of each compartment. Thus, the mitochondria, which produce most of the cell's ATP, contain all of the enzymes needed to carry out the tricarboxylic acid cycle and oxidative phosphorylation. Similarly, the degradative enzymes needed for the intracellular digestion of unwanted macromolecules are confined to the lysosomes.

The relative volumes occupied by some cellular compartments in a typical liver cell		
Cellular compartment	Percent of total cell volume	Approximate number per cell
Cytosol	54	1
Mitochondrion	22	1,700
Endoplasmic reticulum plus golgi apparatus	15	1
Nucleus	6	1
Lysosome	1	300

It is clear from this functional segregation that the many different proteins specified by the genes in the cell nucleus must be transported to the compartment where they will be used. Not surprisingly, the cell contains an extensive membrane-bound system devoted to maintaining just this intracellular order. The system serves as a post office, guaranteeing the proper routing of newly synthesized macromolecules to their proper destinations.

All proteins are synthesized on ribosomes located in the cytosol. As soon as the first portion of the amino acid sequence of a protein emerges from the ribosome, it is inspected for the presence of a short "endoplasmic reticulum (ER) signal sequence." Those ribosomes making proteins with such a sequence are transported to the surface of the ER membrane, where they complete their synthesis; the proteins made on these ribosomes are immediately transferred through the ER membrane to the inside of the ER compartment. Proteins lacking the ER signal sequence remain in the cytosol and are released from the ribosomes when their synthesis is completed. This chemical decision process places some newly completed protein chains in the cytosol and others within an extensive membrane-bounded compartment in the cytoplasm, representing the first step in intracellular protein sorting.

The newly made proteins in both cell compartments are then sorted further according to additional signal sequences that they contain. Some of the proteins in the cytosol remain there, while others go to the surface of mitochondria or (in plant cells) chloroplasts, where they are transferred through the membranes into the organelles. Subsignals on each of these proteins then designate exactly where in the organelle the protein belongs. The proteins initially sorted into the ER have an even wider range of destinations. Some of them remain in the ER, where they function as part of the organelle. Most enter transport vesicles and pass to the Golgi apparatus, separate membrane-bounded organelles that contain at least three subcompartments. Some of the proteins are retained in the subcompartments of the Golgi, where they are utilized for functions peculiar to that organelle. Most eventually enter vesicles that leave the Golgi for other cellular destinations such as the cell membrane, lysosomes, or special secretory vesicles.

Intercellular Communication

Formation of a multicellular organism starts with a small collection of similar cells in an embryo and proceeds by continuous cell division and specialization to produce an entire community of cooperating cells, each with its own role in the life of the organism. Through cell cooperation, the organism becomes much more than the sum of its component parts.

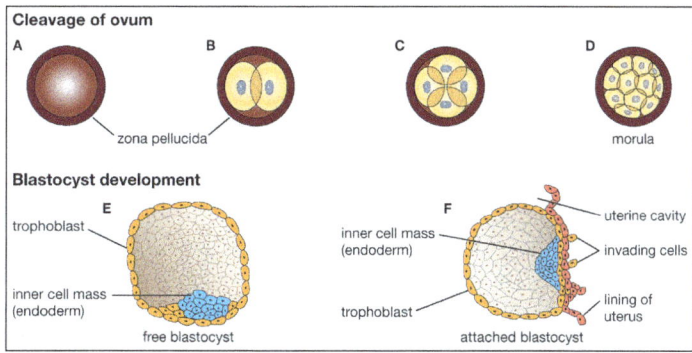

The ovum contains a small collection of cells in the early stages of human development. As cells divide (A–D), they are separated into different regions of the ovum. Each region of the ovum transmits a unique set of chemical signals to nearby cells. Thus, the signals detected by one cell differ from those detected by its neighbour cells. In this process, known as cell determination, cells are individually programmed to direct them toward development into different cell types.

A fertilized egg multiplies and produces a whole family of daughter cells, each of which adopts a structure and function according to its position in the entire assembly. All of the daughter cells contain the same chromosomes and therefore the same genetic information. Despite this common inheritance, different types of cells behave differently and have different structures. In order for this to be the case, they must express different sets of genes, so that they produce different proteins despite their identical embryological ancestors.

During the development of an embryo, it is not sufficient for all the cell types found in the fully developed individual simply to be created. Each cell type must form in the right place at the right time and in the correct proportion; otherwise, there would be a jumble of randomly assorted cells in no way resembling an organism. The orderly development of an organism depends on a process called cell determination, in which initially identical cells become committed to different pathways of development. A fundamental part of cell determination is the ability of cells to detect different chemicals within different regions of the embryo. The chemical signals detected by one cell may be different from the signals detected by its neighbour cells. The signals that a cell detects activate a set of genes that tell the cell to differentiate in ways appropriate for its position within the embryo. The set of genes activated in one cell differs from the set of genes activated in the cells around it. The process of cell determination requires an elaborate system of cell-to-cell communication in early embryos.

The Cell Membrane

A thin membrane, typically between 4 and 10 nanometers (nm; 1 nm = 10^{-9} metre) in thickness, surrounds every living cell, delimiting the cell from the environment around it. Enclosed by this cell membrane (also known as the plasma membrane) are the cell's constituents, often large, water-soluble, highly charged molecules such as proteins, nucleic acids, carbohydrates, and substances involved in cellular metabolism. Outside the cell, in the surrounding water-based environment, are ions, acids, and alkalis that are toxic to the cell, as well as nutrients that the cell must absorb in order to live and grow. The cell membrane, therefore, has two functions: first, to be a barrier keeping the constituents of the cell in and unwanted substances out and, second, to be a gate allowing transport into the cell of essential nutrients and movement from the cell of waste products.

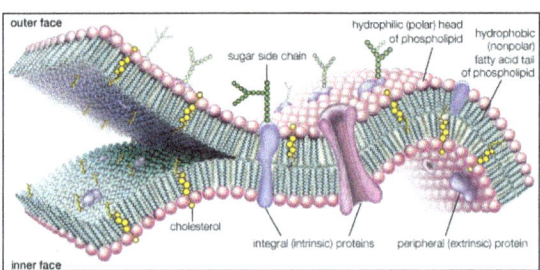

Molecular view of the cell membraneIntrinsic proteins penetrate and bind tightly to the lipid bilayer, which is made up largely of phospholipids and cholesterol and which typically is between 4 and 10 nanometers (nm; 1 nm = 10^{-9} metre) in thickness. Extrinsic proteins are loosely bound to the hydrophilic (polar) surfaces, which face the watery medium both inside and outside the cell. Some intrinsic proteins present sugar side chains on the cell's outer surface.

Chemical Composition and Membrane Structure

Most current knowledge about the biochemical constituents of cell membranes originates in studies of red blood cells. The chief advantage of these cells for experimental purposes is that they may be obtained easily in large amounts and that they have no internal membranous organelles to interfere with study of their cell membranes. Careful studies of these and other cell types have shown that all membranes are composed of proteins and fatty-acid-based lipids. Membranes actively involved in metabolism contain a higher proportion of protein; thus, the membrane of the mitochondrion, the most rapidly metabolizing organelle of the cell, contains as much as 75 percent protein, while the membrane of the Schwann cell, which forms an insulating sheath around many nerve cells, has as little as 20 percent protein.

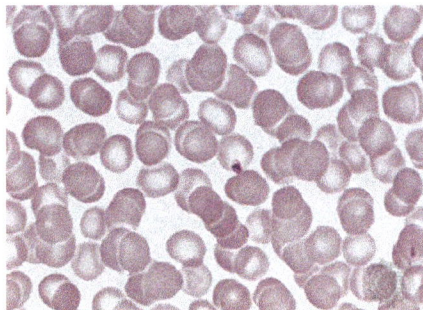
Human red blood cells (erythrocytes).

Membrane Lipids

Membrane lipids are principally of two types, phospholipids and sterols (generally cholesterol). Both types share the defining characteristic of lipids—they dissolve readily in organic solvents—but in addition they both have a region that is attracted to and soluble in water. This "amphiphilic" property (having a dual attraction; i.e., containing both a lipid-soluble and a water-soluble region) is basic to the role of lipids as building blocks of cellular membranes. Phospholipid molecules have a head (often of glycerol) to which are attached two long fatty acid chains that look much like tails. These tails are repelled by water and dissolve readily in organic solvents, giving the molecule its lipid character. To another part of the head is attached a phosphoryl group with a negative electrical charge; to this group in turn is attached another group with a positive or neutral charge. This portion of the phospholipid dissolves in water, thereby completing the molecule's amphiphilic character. In contrast, sterols have a complex hydrocarbon ring structure as the lipid-soluble region and a hydroxyl grouping as the water-soluble region.

General structural formula of a glycerophospholipid. The composition of the specific molecule depends on the chemical group (designated R3 in the diagram) linked to the phosphate and glycerol "head" and also on the lengths of the fatty acid "tails" (R1 and R2).

When dry phospholipids, or a mixture of such phospholipids and cholesterol, are immersed in water under laboratory conditions, they spontaneously form globular structures called liposomes. Investigation of the liposomes shows them to be made of concentric spheres, one sphere inside of another and each forming half of a bilayered wall. A bilayer is composed of two sheets of phospholipid molecules with all of the molecules of each sheet aligned in the same direction. In a water medium, the phospholipids of the two sheets align so that their water-repellent, lipid-soluble tails are turned and loosely bonded to the tails of the molecules on the other sheet. The water-soluble heads turn outward into the water, to which they are chemically attracted. In this way, the two sheets form a fluid, sandwichlike structure, with the fatty acid chains in the middle mingling in an organic medium while sealing out the water medium.

This type of lipid bilayer, formed by the self-assembly of lipid molecules, is the basic structure of the cell membrane. It is the most stable thermodynamic structure that a phospholipid-water mixture can take up: the fatty acid portion of each molecule dissolved in the organic phase formed by the identical regions of the other molecules and the water-attractive regions surrounded by water and facing away from the fatty acid regions. The chemical affinity of each region of the amphiphilic molecule is thus satisfied in the bilayer structure.

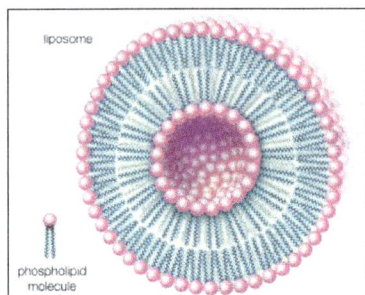

Phospholipids can be used to form artificial structures called liposomes, which are double-walled, hollow spheres useful for encapsulating other molecules such as pharmaceutical drugs.

Phospholipid molecules, like molecules of many lipids, are composed of a hydrophilic "head" and one or more hydrophobic "tails." In a water medium, the molecules form a lipid bilayer, or two-layered sheet, in which the heads are turned toward the watery medium and the tails are sheltered inside, away from the water. This bilayer is the basis of the membranes of living cells.

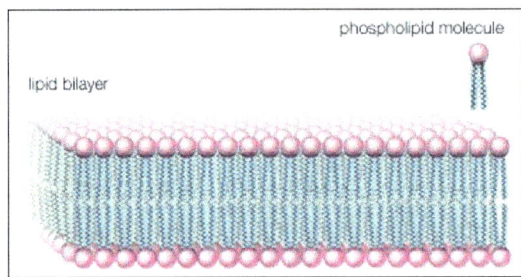

Membrane Proteins

Membrane proteins are also of two general types. One type, called the extrinsic proteins, is loosely attached by ionic bonds or calcium bridges to the electrically charged phosphoryl surface of the bilayer. They can also attach to the second type of protein, called the intrinsic proteins. The intrinsic proteins, as their name implies, are firmly embedded within the phospholipid bilayer. Almost all intrinsic proteins contain special amino acid sequences, generally about 20- to 24-amino acids long, that extend through the internal regions of the cell membrane.

Most intrinsic and extrinsic proteins bear on their outer surfaces side chains of complex sugars, which extend into the aqueous environment around the cell. For this reason, these proteins are often referred to as glycoproteins. Some glycoproteins are involved in cell-to-cell recognition.

Membrane Fluidity

One of the triumphs of cell biology during the decade from 1965 to 1975 was the recognition of the cell membrane as a fluid collection of amphiphilic molecules. This array of proteins, sterols, and phospholipids is organized into a liquid crystal, a structure that lends itself readily to rapid cell growth. Measurements of the membrane's viscosity show it as a fluid one hundred times as viscous as water, similar to a thin oil. The phospholipid molecules diffuse readily in the plane of the bilayer. Many of the membrane's proteins also have this freedom of movement, but some are fixed in the membrane by interaction with the cell's cytoskeleton. Newly synthesized phospholipids insert themselves easily into the existing cell membrane. Intrinsic proteins are inserted during their synthesis on ribosomes bound to the endoplasmic reticulum, whereas extrinsic proteins found on the internal surface of the cell membrane are synthesized on free, or unattached, ribosomes, liberated into the cytoplasm, and then brought to the membrane.

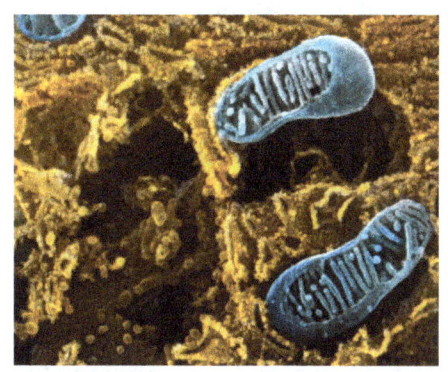

Endoplasmic reticulum; organelleA scanning electron micrograph of a pancreatic acinar cell, showing mitochondria (blue), rough endoplasmic reticulum (yellow; ribosomes appear as small dots), and Golgi apparatus (gray, at centre and lower left).

Transport across the Membrane

The chemical structure of the cell membrane makes it remarkably flexible, the ideal boundary for rapidly growing and dividing cells. Yet the membrane is also a formidable barrier, allowing some dissolved substances, or solutes, to pass while blocking others. Lipid-soluble molecules and some small molecules can permeate the membrane, but the lipid bilayer effectively repels the many large, water-soluble molecules and electrically charged ions that the cell must import or export in order to live. Transport of these vital substances is carried out by certain classes of intrinsic proteins that form a variety of transport systems: some are open channels, which allow ions to diffuse directly into the cell; others are "facilitators," which, through a little-understood chemical transformation, help solutes diffuse past the lipid screen; yet others are "pumps," which force solutes through the membrane when they are not concentrated enough to diffuse spontaneously. Particles too large to be diffused or pumped are often swallowed or disgorged whole by an opening and closing of the membrane.

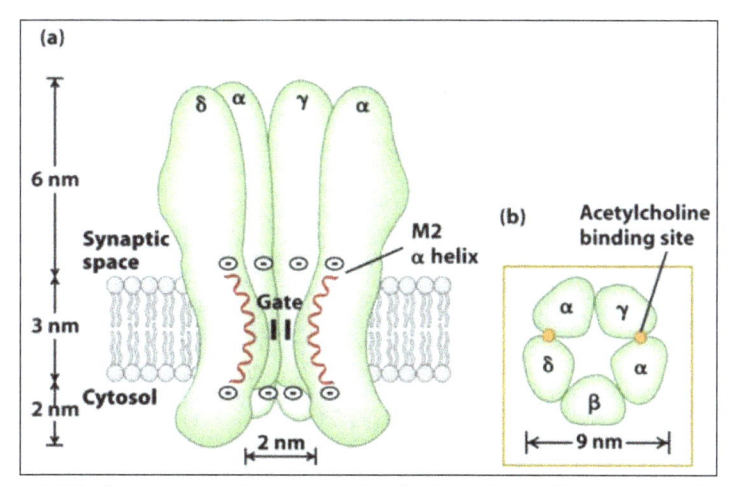

The nicotinic acetylcholine receptor is an example of a ligand-gated ion channel. It is composed of five subunits arranged symmetrically around a central conducting pore. Upon binding acetylcholine, the channel opens and allows diffusion of sodium (Na^+) and potassium (K^+) ions through the conducting pore.

Behind this movement of solutes across the cell membrane is the principle of diffusion. According to this principle, a dissolved substance diffuses down a concentration gradient; that is, given no energy from an outside source, it moves from a place where its concentration is high to a place where its concentration is low. Diffusion continues down this gradually decreasing gradient until a state of equilibrium is reached, at which point there is an equal concentration in both places and an equal, random diffusion in both directions.

A solute at high concentration is at high free energy; that is, it is capable of doing more "work" (the work being that of diffusion) than a solute at low concentration. In performing the work of diffusion, the solute loses free energy, so that, when it reaches equilibrium at a lower concentration, it is unable to return spontaneously (under its own energy) to its former high concentration. However, by the addition of energy from an outside source (through the work of an ion pump, for example), the solute may be returned to its former concentration and state of high free energy. This "coupling" of work processes is, in effect, a transferal of free energy from the pump to the solute, which is then able to repeat the work of diffusion.

For most substances of biological interest, the concentrations inside and outside the cell are different, creating concentration gradients down which the solutes spontaneously diffuse, provided they can permeate the lipid bilayer. Membrane channels and diffusion facilitators bring them through the membrane by passive transport; that is, the changes that the proteins undergo in order to facilitate diffusion are powered by the diffusing solutes themselves. For the healthy functioning of the cell, certain solutes must remain at different concentrations on each side of the membrane; if through diffusion they approach equilibrium, they must be pumped back up their gradients by the process of active transport. Those membrane proteins serving as pumps accomplish this by coupling the energy required for transport to the energy produced by cell metabolism or by the diffusion of other solutes.

Permeation

Permeation is the diffusion, through a barrier, of a substance in solution. The rates at which biologically important molecules cross the cell membrane through permeation vary over an enormous range. Proteins and sugar polymers do not permeate at all; in contrast, water and alcohols permeate most membranes in less than a second. This variation, caused by the lipid bilayer, gives the membrane its characteristic permeability. Permeability is measured as the rate at which a particular substance in solution crosses the membrane.

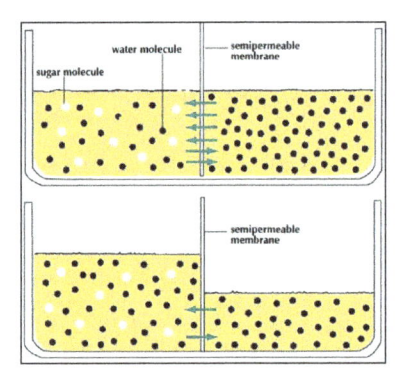

The principle of permeation can be illustrated by differences in the diffusion of sugar and water through a membrane. Large sugar molecules in the solution cannot pass through the membrane into the water (top). In contrast, small water molecules easily diffuse through the membrane (bottom). The ability of water to readily cross membranes is vital for establishing equilibrium.

For all cell membranes that have been studied in the laboratory, permeability increases in parallel with the permeant's ability to dissolve in organic solvents. The consistency of this parallel has led researchers to conclude that permeability is a function of the fatty acid interior of the lipid bilayer, rather than its phosphoryl exterior. This property of dissolving in organic solvents rather than water is given a unit of measure called the partition coefficient. The greater the solubility of a substance, the higher its partition coefficient, and the higher the partition coefficient, the higher the permeability of the membrane to that particular substance. For example, the water solubility of hydroxyl, carboxyl, and amino groups reduces their solubility in organic solvents and, hence, their partition coefficients. Cell membranes have been observed to have low permeability toward these groups. In contrast, lipid-soluble methyl residues and hydrocarbon rings, which have high partition coefficients, penetrate cell membranes more easily—a property useful in designing chemotherapeutic and pharmacological drugs.

For two molecules of the same partition coefficient, the one of greater molecular weight, or size, will in general cross the membrane more slowly. In fact, even molecules with very low partition coefficients can penetrate the membrane if they are small enough. Water, for example, is insoluble in organic solvents, yet it permeates cell membranes because of the small size of its molecules. The size selectivity of the lipid bilayer is a result of its being not a simple fluid, the molecules of which move around and past a diffusing molecule, but an organized matrix, a kind of fixed grate, composed of the fatty acid chains of the phospholipids through which the diffusing molecule must fit.

Many substances do not actually cross the cell membrane through permeation of the lipid bilayer. Some electrically charged ions, for example, are repelled by organic solvents and therefore cross cell membranes with great difficulty, if at all. In these cases special holes in the membrane, called channels, allow specific ions and small molecules to diffuse directly through the bilayer.

Membrane Channels

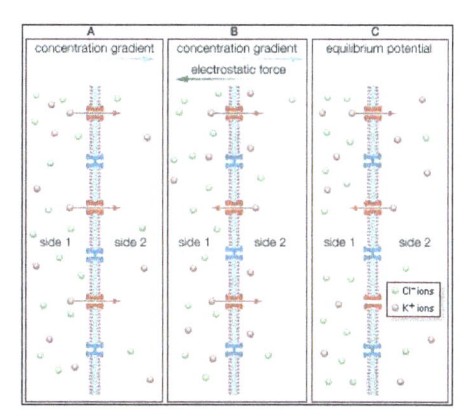

Diffusion of ions across a semipermeable membrane: (A) A high concentration of KCl is placed on side 1, opposite a semipermeable membrane from a low concentration. The membrane allows only K+ to diffuse, thereby establishing an electrical potential difference across the membrane. (B) The separation of charge creates an electrostatic voltage force, which draws some K+ back to side 1. (C) At equilibrium, there is no net flux of K+ in either direction. Side 1, with the higher concentration of KCl, has a negative charge compared with side 2.

Biophysicists measuring the electric current passing through cell membranes have found that, in general, cell membranes have a vastly greater electrical conductance than does a membrane bilayer composed only of phospholipids and sterols. This greater conductance is thought to be conferred by the cell membrane's proteins. A current flowing across a membrane often appears on a recording instrument as a series of bursts of various heights. These bursts represent current flowing through open channels, which are merely holes formed by intrinsic proteins traversing the lipid bilayer. No significant current flows through the membrane when no channel is open; multiple bursts are recorded when more than one channel is open.

A rich variety of channels has been isolated and analyzed from a wide range of cell membranes. Invariably intrinsic proteins, they contain numerous amino acid sequences that traverse the membrane, clearly forming a specific hole, or pore. Certain channels open and close spontaneously. Some are gated, or opened, by the chemical action of a signaling substance such as calcium, acetylcholine, or glycine, whereas others are gated by changes in the electrical potential across the membrane. Channels may possess a narrow specificity, allowing passage of only potassium or sodium, or a broad specificity, allowing passage of all positively charged ions (cations) or of all negatively charged ions (anions). There are channels called gap junctions that allow the passage of molecules between pairs of cells.

The gating of channels with a capacity for ion transport is the basis of the many nerve-nerve, nerve-muscle, and nerve-gland interactions underlying neurobiological behaviour. These actions depend on the electric potential of the cell membrane, which varies with the prevailing constituents in the cell's environment. For example, if a channel that admits only potassium ions is present in a membrane separating two different potassium chloride solutions, the positively charged potassium ions tend to flow down their concentration gradient through the channel. The negatively charged chloride ions remain behind. This separation of electric charges sets up an electric potential across the membrane called the diffusion potential. The size of this potential depends on, among other factors, the difference in concentrations of the permeating ion across the membrane. The cell membrane in general contains the channels of widely different ion specificities, each channel contributing to the overall membrane potential according to the permeability and concentration ratio of the ion passing through it. Since the channels are often gated, the membrane's potential is determined by which channels are open; this in turn depends on the concentrations of signaling molecules and may change with time according to the membrane potential itself.

Most cells have about a tenfold higher concentration of sodium ions outside than inside and a reverse concentration ratio of potassium ions. Free calcium ions can be 10,000 times more concentrated outside the cell than inside. Thus, sodium-, potassium-, and calcium-selective membrane channels, by allowing the diffusion of those ions past the cell membrane and causing fluctuations in the membrane's electric potential, frequently serve as transmitters of signals from nerve cells. Ion diffusion threatens to alter the concentration of ions necessary for the cell to function. The proper distribution of ions is restored by the action of ion pumps.

Facilitated Diffusion

Many water-soluble molecules that cannot penetrate the lipid bilayer are too large to fit through open channels. In this category are sugars and amino acids. Some ions too do not diffuse through channels. These vital substances enter and leave the cell through the action of membrane transporters, which, like channels, are intrinsic proteins that traverse the cell membrane. Unlike channels, transporter molecules do not simply open holes in the membrane. Rather, they present sites on one side of the membrane to which molecules bind through chemical attraction. The binding site is highly specific, often fitting the atomic structure of only one type of molecule. When the molecule has attached to the binding site, then, in a process not fully understood, the transporter brings it through the membrane and releases it on the other side.

This action is considered a type of diffusion because the transported molecules move down their concentration gradients, from high concentration to low. To activate the action of the transporter, no other energy is needed than that of the chemical binding of the transported molecules. This action upon the transporter is similar to catalysis, except that the molecules (in this context called substrates) catalyze not a chemical reaction but their own translocation across the cell membrane. Two such substrates are glucose and the bicarbonate ion.

The Glucose Transporter

This sugar-specific transport system enables half of the glucose present inside the cell to leave within four seconds at normal body temperature. The glucose transporter is clearly not a simple membrane channel. First, unlike a channel, it does not select its permeants by size, as one type of glucose is observed to move through the system a thousand times faster than its identically sized optical isomer. Second, it operates much more slowly than do most channels, moving only 1,000 molecules per second while a channel moves 1,000,000 ions. The most important difference between a membrane channel and the glucose transporter is the conformational change that the transporter undergoes while moving glucose across the membrane. Alternating between two conformations, it moves its glucose-binding site from one side of the membrane to the other. By "flipping" between its two conformational states, the transporter facilitates the diffusion of glucose; that is, it enables glucose to avoid the barrier of the cell membrane while moving spontaneously down its concentration gradient. When the concentration reaches equilibrium, net movement of glucose ceases.

A facilitated diffusion system for glucose is present in many cell types. Similar systems transporting a wide range of other substrates (e.g., different sugars, amino acids, nucleosides, and ions) are also present.

The Anion Transporter

The best-studied of the facilitated diffusion systems is that which catalyzes the exchange of anions across the red blood cell membrane. The exchange of hydroxyl for bicarbonate ions, each ion simultaneously being moved down its concentration gradient in opposite directions by the same transport molecule, is of great importance in enhancing the blood's capacity to carry carbon dioxide from tissues to the lungs. The exchange molecule for these anions is the major intrinsic protein of red blood cells; one million of them are present on each cell, the polypeptide chain of each molecule traversing the membrane at least six times.

Secondary Active Transport

In some cases the problem of forcing a substrate up its concentration gradient is solved by coupling that upward movement to the downward flow of another substrate. In this way the energy-expending diffusion of the driving substrate powers the energy-absorbing movement of the driven substrate from low concentration to high. Because this type of active transport is not powered directly by the energy released in cell metabolism, it is called secondary.

There are two kinds of secondary active transport: counter-transport, in which the two substrates cross the membrane in opposite directions, and cotransport, in which they cross in the same direction.

Counter-transport

An example of this system (also called antiport) begins with the sugar transporter described above. There are equal concentrations of glucose on both sides of the cell. A high concentration of galactose is then added outside the cell. Galactose competes with glucose for binding sites on the transport protein, so that mostly galactose—and a little glucose—enter the cell. The transporter itself, undergoing a conformational change, presents its binding sites for sugar at the inner face of the membrane. Here, at least transiently, glucose is in excess of galactose; it binds to the transporter and leaves the cell as the transporter switches back to its original conformation. Thus, glucose is pumped out of the cell against its gradient in exchange for the galactose riding into the cell down its own gradient.

Many counter-transport systems operate across the cell membranes of the body. A well-studied system (present in red blood cells, nerve cells, and muscle cells) pumps one calcium ion out of the cell in exchange for two or three sodium ions. This system helps maintain the low calcium concentration required for effective cellular activity. A different system, present in kidney cells, counter-transports hydrogen ions and sodium ions in a one-for-one ratio. This is important in stabilizing acidity by transporting hydrogen ions out of the body as needed.

Co-transport

In co-transport (sometimes called symport) two species of substrate, generally an ion and another molecule or ion, must bind simultaneously to the transporter before its conformational change can take place. As the driving substrate is transported down its concentration gradient, it drags with it the driven substrate, which is forced to move up its concentration gradient. The transporter must be able to undergo a conformational change when not bound to either substrate, so as to complete the cycle and return the binding sites to the side from which driving and driven substrates both move.

Sodium ions are usually the driving substrates in the co-transport systems of animal cells, which maintain high concentrations of these ions through primary active transport. The driven substrates include a variety of sugars, amino acids, and other ions. During the absorption of nutrients, for example, sugars and amino acids are removed from the intestine by co-transport with sodium ions. After passing across the glomerular filter in the kidney, these substrates are returned to the body by the same system. Plant and bacterial cells usually use hydrogen ions as the driving substrate; sugars and amino acids are the most common driven substrates. When the bacterium Escherichia coli must metabolize lactose, it co-transports hydrogen ions with lactose (which can reach a concentration 1,000 times higher than that outside the cell).

Primary Active Transport

The Sodium-potassium Pump

Human red blood cells contain a high concentration of potassium and a low concentration of sodium, yet the plasma bathing the cells is high in sodium and low in potassium. When whole blood is stored cold under laboratory conditions, the cells lose potassium and gain sodium until the concentrations across the membrane for both ions are at equilibrium. When the cells are restored to body temperature and given appropriate nutrition, they extrude sodium and take up potassium, transporting both ions against their respective gradients until the previous high concentrations are reached. This ion pumping is linked directly to the hydrolysis of adenosine triphosphate (ATP), the cell's repository of metabolic energy. For every molecule of ATP split, three ions of sodium are pumped out of the cell and two of potassium are pumped in.

An enzyme called sodium-potassium-activated ATPase has been shown to be the sodium-potassium pump, the protein that transports the ions across the cell membrane while splitting ATP. Widely distributed in the animal kingdom and always associated with the cell membrane, this ATPase is found at high concentration in cells that pump large amounts of sodium (e.g., in mammalian kidneys, in salt-secreting glands of marine birds, and in the electric organs of eels). The enzyme, an intrinsic protein, exists in two major conformations whose interconversion is driven by the splitting of ATP or by changes in the transmembrane flows of sodium and potassium. When only sodium is present in the cell, the inorganic phosphate split from ATP during hydrolysis is transferred to the enzyme. Release of the chemically bound phosphate from the enzyme is catalyzed by potassium. Thus, the complete action of ATP splitting has been demonstrated to require both sodium (to catalyze the transfer of the phosphate to the enzyme) and potassium (to catalyze the release of the phosphate and free the enzyme for a further cycle of ATP splitting). Apparently, only after sodium has catalyzed the transferal of the phosphate to the enzyme can it be transported from the cell. Similarly, only after potassium has released the phosphate from the enzyme can it be transported into the cell. This overall reaction, completing the cycle of conformational changes in the enzyme, involves a strict coupling of the splitting of ATP with the pumping of sodium and potassium. It is this coupling that creates primary active transport.

The sodium-potassium pump extrudes one net positive charge during each cycle of ATP splitting. This flow of current induces an electric potential across the membrane that adds to the potentials brought about by the diffusion of ions through gated channels. The pump's contribution to the overall potential is important in certain specialized nerve cells.

Calcium Pumps

Many animal cells can perform a primary active transport of calcium out of the cell, developing a 10,000-fold gradient of that ion. Calcium-activated ATPases have been isolated and shown to be intrinsic proteins straddling the membrane and undergoing conformational changes similar to those of the sodium-potassium-activated ATPase. When a rise in the concentration of cellular calcium results from the opening of calcium-selective channels, the membrane's calcium pumps restore the low concentration.

Hydrogen Ion Pumps

Hydrochloric acid is produced in the stomach by the active transport of hydrogen ions from the blood across the stomach lining, or gastric mucosa. Hydrogen concentration gradients of nearly one million can be achieved by a hydrogen-potassium-activated ATP-splitting intrinsic protein in the cells lining the stomach. Apart from its specific ion requirements, the properties of this enzyme are remarkably similar to those of the sodium-potassium-activated enzyme and the calcium-activated enzyme. Other hydrogen-pumping ATP-splitting primary active transporters occur in intracellular organelles, in bacteria, and in plant cells. The steep gradient of hydrogen ions represents a store of energy that can be harnessed to the accumulation of nutrients or, in the case of bacterial flagella, to the powering of cell movement.

Transport of Particles

In bringing about transmembrane movements of large molecules, the cell membrane itself undergoes concerted movements during which part of the fluid medium outside of the cell is internalized (endocytosis) or part of the cell's internal medium is externalized (exocytosis). These movements involve a fusion between membrane surfaces, followed by the re-formation of intact membranes.

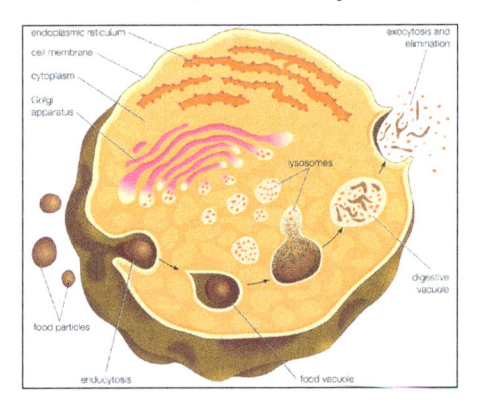

Endocytosis and exocytosis are fundamental to the process of intracellular digestion. Food particles are taken into the cell via endocytosis into a vacuole. Lysosomes attach to the vacuole and release digestive enzymes to extract nutrients. The leftover waste products of digestion are carried to the plasma membrane by the vacuole and eliminated through the process of exocytosis.

Endocytosis

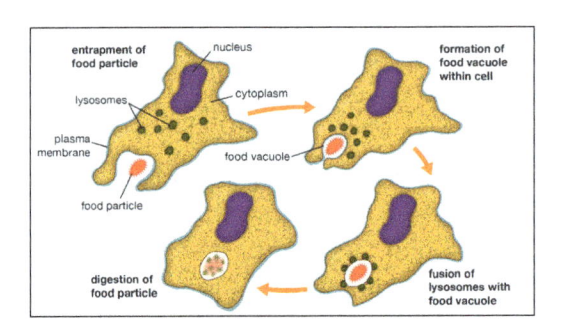

The process by which cells engulf solid matter is called phagocytosis. There are four essential steps in phagocytosis: (1) the plasma membrane entraps the food particle, (2) a vacuole forms within the cell to contain the food particle, (3) lysosomes fuse with the food vacuole, and (4) enzymes of the lysosomes digest the food particle.

In this process the cell membrane engulfs portions of the external medium, forms an almost complete sphere around it, and then draws the membrane-bounded vesicle, called an endosome, into the cell. Several types of endocytosis have been distinguished: in pinocytosis, the vesicles are small and contain fluid; in phagocytosis, the vesicles are larger and contain solid matter; and in receptor-mediated endocytosis, material binds to a specific receptor on the external face of the cell membrane, triggering the process by which it is engulfed. Cholesterol enters cells by the last route.

Exocytosis

In exocytosis, material synthesized within the cell that has been packaged into membrane-bound vesicles is exported from the cell following the fusion of the vesicles with the external cell membrane. The materials so exported are cell-specific protein products, neurotransmitters, and a variety of other molecules.

Internal Membranes

The presence of internal membranes distinguishes eukaryotic cells (cells with a nucleus) from prokaryotic cells (those without a nucleus). Prokaryotic cells are small (one to five micrometres in length) and contain only a single cell membrane; metabolic functions are often confined to different patches of the membrane rather than to areas in the body of the cell. Typical eukaryotic cells, by contrast, are much larger, the cell membrane constituting only 10 percent or less of the total cellular membrane. Metabolic functions in these cells are carried out in the organelles, compartments sequestered from the cell body, or cytoplasm, by internal membranes.

This topic discusses internal membranes as structural and functional components in the organelles and vesicles of eukaryotic cells. The principal organelles—the nucleus, mitochondrion, and (in plants) chloroplast—are discussed elsewhere. Of the remaining organelles, the lysosomes, peroxisomes, and (in plants) glyoxysomes enclose extremely reactive by-products and enzymes. Internal membranes form the mazelike endoplasmic reticulum, where cell membrane proteins and lipids are synthesized, and they also form the stacks of flattened sacs called the Golgi apparatus, which is associated with the transport and modification of lipids, proteins, and carbohydrates. Finally, internal cell membranes can form storage and transport vesicles and the vacuoles of plant cells. Each membrane structure has its own distinct composition of proteins and lipids enabling it to carry out unique functions.

General Functions and Characteristics

Like the cell membrane, membranes of some organelles contain transport proteins, or permeases, that allow chemical communication between organelles. Permeases in the lysosomal membrane, for example, allow amino acids generated inside the lysosome to cross into the cytoplasm, where they can be used for the synthesis of new proteins. Communication between organelles is also achieved by the membrane budding processes of endocytosis and exocytosis, which are essentially the same as in the cell membrane. On the other hand, the biosynthetic and degradative processes taking place in different organelles may require conditions greatly different from those of other organelles or of the cytosol (the fluid part of the cell surrounding the organelles). Internal membranes maintain these different conditions by isolating them from one another. For example, the internal space of lysosomes is much more acidic than that of the cytosol—pH 5 as opposed to pH 7—and is maintained by specific proton-pumping transport proteins in the lysosome membrane.

Another function of organelles is to prevent competing enzymatic reactions from interfering with one another. For instance, essential proteins are synthesized on the rough endoplasmic reticulum and in the cytosol, while unwanted proteins are broken down in the lysosomes and also, to some extent, in the cytosol. Similarly, fatty acids are made in the cytosol and then either broken down in the mitochondria for the synthesis of ATP or degraded in the peroxisomes with concomitant generation of heat. These processes must be kept isolated. Organelle membranes also prevent potentially lethal by-products or enzymes from attacking sensitive molecules in other regions of the cell by sequestering such degradative activities in their respective membrane-bounded compartments.

The internal membranes of eukaryotic cells differ both structurally and chemically from the outer cell membrane. Like the outer membrane, they are constructed of a phospholipid bilayer into which are embedded, or bound, specific membrane proteins. The three major lipids forming the outer membrane—phospholipids, cholesterol, and glycolipids—are also found in the internal membranes, but in different concentrations. Phospholipid is the primary lipid forming all cellular membranes. Cholesterol, which contributes to the fluidity and stability of all membranes, is found in internal membranes at about 25 percent of the concentration in the outer membrane. Glycolipids are found only as trace components of internal membranes, whereas they constitute approximately 5 percent of the outer membrane lipid.

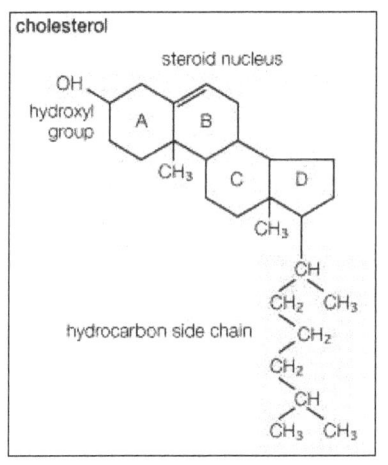

Nucleus; animal cell.

Cellular Organelles and their Membranes

The Vacuole

Most plant cells contain one or more membrane-bound vesicles called vacuoles. Within the vacuole is the cell sap, a water solution of salts and sugars kept at high concentration by the active transport of ions through permeases in the vacuole membrane. Proton pumps also maintain high concentrations of protons in the vacuole interior. These high concentrations cause the entry, via osmosis, of water into the vacuole, which in turn expands the vacuole and generates a hydrostatic pressure, called turgor, that presses the cell membrane against the cell wall. Turgor is the cause of rigidity in living plant tissue.

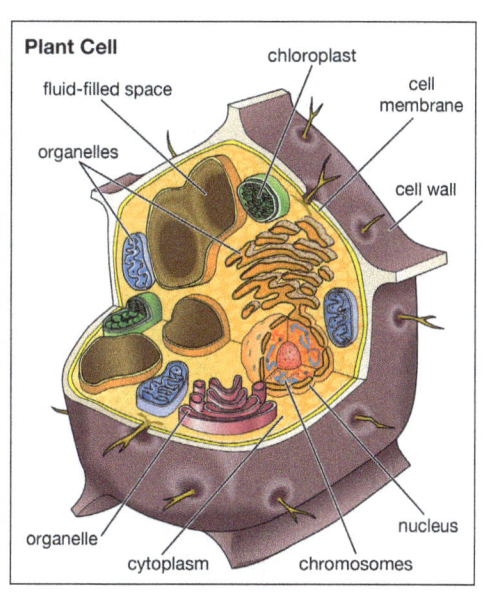

Plant cells contain membrane-bound organelles, including fluid-filled spaces, called vacuoles, that play an important role in maintaining the rigidity of a plant.

In the mature plant cell, as much as 90 percent of cell volume may be taken up by a single vacuole; immature cells typically contain several smaller vacuoles.

The Lysosome

Potentially dangerous hydrolytic enzymes functioning in acidic conditions (pH 5) are segregated in the lysosomes to protect the other components of the cell from random destruction. Lysosomes are bound by a single phospholipid bilayer membrane. They vary in size and are formed by the fusion of Golgi-derived vesicles with endosomes derived from the cell surface. Enzymes known to be present in the lysosomes include hydrolases that degrade proteins, nucleic acids, lipids, glycolipids, and glycoproteins. Hydrolases are most active in the acidity maintained in the lysosomes. After the material is broken down, lipids and amino acids are transported across the lysosomal membrane by permeases for use in biosynthesis. The remaining debris generally stays within the lysosome and is called a residual body.

Microbodies

Microbodies are roughly spherical in shape, bound by a single membrane, and are usually 0.5 to 1 micrometre in diameter. There are several types, by far the most common of which is the peroxisome. Peroxisomes derive their name from hydrogen peroxide, a reactive intermediate in the process of molecular breakdown that occurs in the microbody. Peroxisomes contain type II oxidases, which are enzymes that use molecular oxygen in reactions to oxidize organic molecules. A product of these reactions is hydrogen peroxide, which is further metabolized into water and oxygen by the enzyme catalase, a predominant constituent of peroxisomes. In addition, peroxisomes contain other enzyme systems that degrade various lipids.

The plant glyoxysome is a peroxisome that also contains the enzymes of the glyoxylate cycle, which is crucial to the conversion of fat into carbohydrate.

Endoplasmic Reticulum

The endoplasmic reticulum (ER) is a system of membranous cisternae (flattened sacs) extending throughout the cytoplasm. Often it constitutes more than half of the total membrane in the cell. This structure was first noted in the late 19th century, when studies of stained cells indicated the presence of some type of extensive cytoplasmic structure, then termed the gastroplasm. The electron microscope made possible the study of the morphology of this organelle in the 1940s, when it was given its present name.

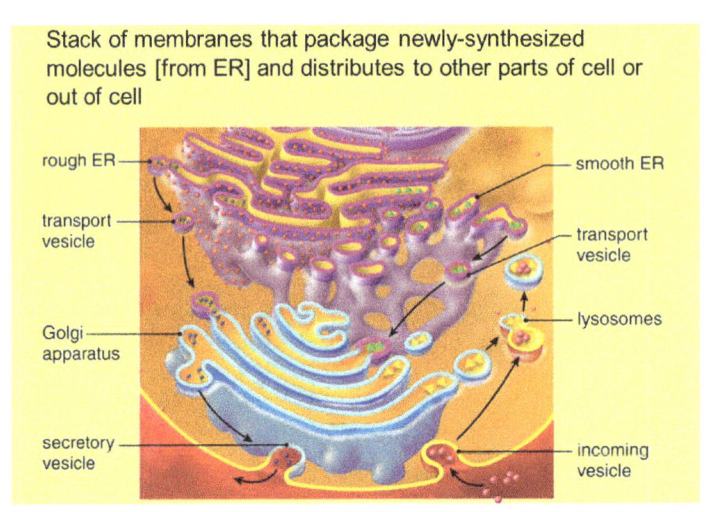

Stack of membranes that package newly-synthesized molecules [from ER] and distributes to other parts of cell or out of cell

- rough ER
- smooth ER
- transport vesicle
- transport vesicle
- Golgi apparatus
- lysosomes
- secretory vesicle
- incoming vesicle

The endoplasmic reticulum (ER) plays a major role in the biosynthesis of proteins. Proteins that are synthesized by ribosomes on the ER are transported into the Golgi apparatus for processing. Some of these proteins will be secreted from the cell, others will be inserted into the plasma membrane, and still others will be inserted into lysosomes.

The endoplasmic reticulum can be classified in two functionally distinct forms, the smooth endoplasmic reticulum (SER) and the rough endoplasmic reticulum (RER). The morphological distinction between the two is the presence of protein-synthesizing particles, called ribosomes, attached to the outer surface of the RER.

The Smooth Endoplasmic Reticulum

The functions of the SER, a meshwork of fine tubular membrane vesicles, vary considerably from cell to cell. One important role is the synthesis of phospholipids and cholesterol, which are major components of the plasma and internal membranes. Phospholipids are formed from fatty acids, glycerol phosphate, and other small water-soluble molecules by enzymes bound to the ER membrane with their active sites facing the cytosol. Some phospholipids remain in the ER membrane, where, catalyzed by specific enzymes within the membranes, they can "flip" from the cytoplasmic side of the bilayer, where they were formed, to the exoplasmic, or inner, side. This process ensures the symmetrical growth of the ER membrane. Other phospholipids are transferred through the cytoplasm to other membranous structures, such as the cell membrane and the mitochondrion, by special phospholipid transfer proteins.

In liver cells, the SER is specialized for the detoxification of a wide variety of compounds produced by metabolic processes. Liver SER contains a number of enzymes called cytochrome P450, which catalyze the breakdown of carcinogens and other organic molecules. In cells of the adrenal glands and gonads, cholesterol is modified in the SER at one stage of its conversion to steroid hormones.

Finally, the SER in muscle cells, known as the sarcoplasmic reticulum, sequesters calcium ions from the cytoplasm. When the muscle is triggered by nerve stimuli, the calcium ions are released, causing muscle contraction.

Rough Endoplasmic Reticulum

The RER is generally a series of connected flattened sacs. It plays a central role in the synthesis and export of proteins and glycoproteins and is best studied in the secretory cells specialized in these functions. The many secretory cells in the human body include liver cells secreting serum proteins such as albumin, endocrine cells secreting peptide hormones such as insulin, salivary gland and pancreatic acinar cells secreting digestive enzymes, mammary gland cells secreting milk proteins, and cartilage cells secreting collagen and proteoglycans.

Ribosomes are particles that synthesize proteins from amino acids. They are composed of four RNA molecules and between 40 and 80 proteins assembled into a large and a small subunit. Ribosomes are either free (i.e., not bound to membranes) in the cytoplasm of the cell or bound to the RER. Lysosomal enzymes, proteins destined for the ER, Golgi, and cell membranes, and proteins to be secreted from the cell are among those synthesized on membrane-bound ribosomes. Fabricated on free ribosomes are proteins remaining in the cytosol and those bound to the internal surface of the outer membrane, as well as those to be incorporated into the nucleus, mitochondria, chloroplasts, peroxisomes, and other organelles. Special features of proteins label them for transport to specific destinations inside or outside of the cell. In 1971 German-born cellular and molecular biologist Günter Blobel and Argentinian-born cellular biologist David Sabatini suggested that the amino-terminal portion of the protein (the first part of the molecule to be made) could act as a "signal sequence." They proposed that such a signal sequence would facilitate the attachment of the growing protein to the ER membrane and lead the protein either into the membrane or through the membrane into the ER lumen (interior).

The signal hypothesis has been substantiated by a large body of experimental evidence. Translation of the blueprint for a specific protein encoded in a messenger RNA molecule begins on a free ribosome. As the growing protein, with the signal sequence at its amino-terminal end, emerges from the ribosome, the sequence binds to a complex of six proteins and one RNA molecule known as the signal recognition particle (SRP). The SRP also binds to the ribosome to halt further formation of the protein. The membrane of the ER contains receptor sites that bind the SRP-ribosome complex to the RER membrane. Upon binding, translation resumes, with the SRP dissociating from the complex and the signal sequence and remainder of the nascent protein threading through the membrane, via a channel called a translocon, into the ER lumen. At that point, the protein is permanently segregated from the cytosol. In most cases, the signal sequence is cleaved from the protein by an enzyme called signal peptidase as it emerges on the luminal surface of the ER membrane. In addition, in a process known as glycosylation, oligosaccharide (complex sugar) chains are often added to the protein to form a glycoprotein. Inside the ER lumen, the protein folds into its characteristic three-dimensional conformation.

Within the lumen, proteins that will be secreted from the cell diffuse into the transitional portion of the ER, a region that is largely free of ribosomes. There the molecules are packaged into small membrane-bounded transport vesicles, which separate from the ER membrane and move through the cytoplasm to a target membrane, usually the Golgi complex. There the transport vesicle membrane fuses with the Golgi membrane, and the contents of the vesicle are delivered into the lumen of the Golgi. This, like all processes of vesicle budding and fusion, preserves the sidedness of the membranes; that is, the cytoplasmic surface of the membrane always faces outward, and the luminal contents are always sequestered from the cytoplasm.

Certain nonsecretory proteins made on the RER remain part of the membrane system of the cell. These membrane proteins have, in addition to the signal sequence, one or more anchor regions composed of lipid-soluble amino acids. The amino acids prevent passage of the protein completely into the ER lumen by anchoring it into the phospholipid bilayer of the ER membrane.

The Golgi Apparatus

The Golgi complex is the site of the modification, completion, and export of secretory proteins and glycoproteins. This organelle, first described by the Italian cytologist Camillo Golgi in 1898, has a characteristic structure composed of five to eight flattened, disk-shaped, membrane-defined cisternae arranged in a stack. Secretory proteins and glycoproteins, cell membrane proteins and glycoproteins, lysosomal proteins, and some glycolipids all pass through the Golgi structure at some point in their maturation. In plant cells, much of the cell wall material passes through the Golgi as well.

The Golgi apparatus itself is structurally polarized, consisting of a "cis" face near the transitional region of the RER, a medial segment, and a "trans" face near the cell membrane. These faces are biochemically distinct, and the enzymatic content of each segment is markedly different. The cis face membranes are generally thinner than the others.

As the secretory proteins move through the Golgi, a number of chemical modifications may transpire. Important among these is the modification of carbohydrate groups. As described above, many secretory proteins are glycosylated in the ER. In the Golgi, specific enzymes modify the oligosaccharide chains of the glycoproteins by removing certain mannose residues and adding other sugars, such as galactose and sialic acid. These enzymes are known collectively as glycosidases and glycosyltransferases. Some secretory proteins will cease to be transported if their carbohydrate groups are modified incorrectly or not permitted to form. In some cases the carbohydrate groups are necessary for the stability or activity of the protein or for targeting the molecule for a specific destination.

Also within the Golgi or secretory vesicles are proteases that cut many secretory proteins at specific amino acid positions. This often results in activation of the secretory protein, an example being the conversion of inactive proinsulin to active insulin by removing a series of amino acids.

Secretory Vesicles

The release of proteins or other molecules from a secretory vesicle is most often stimulated by a nervous or hormonal signal. For example, a nerve cell impulse triggers the fusion of secretory vesicles to the membrane at the nerve terminal, where the vesicles release neurotransmitters into the synaptic cleft (the gap between nerve endings). The action is one of exocytosis: the vesicle and the cell membrane fuse, allowing the proteins and glycoproteins in the vesicle to be released to the cell exterior.

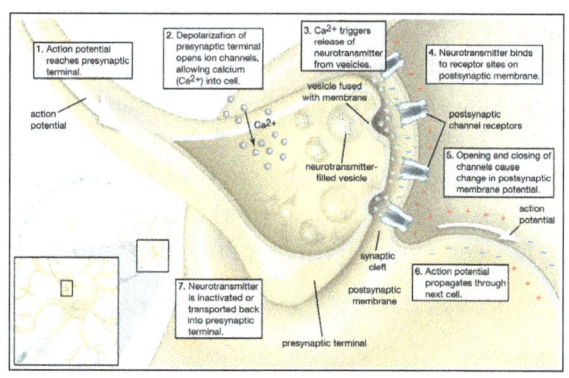

Chemical transmission of a nerve impulse at the synapseThe arrival of the nerve impulse at the presynaptic terminal stimulates the release of neurotransmitter into the synaptic gap. The binding of the neurotransmitter to receptors on the postsynaptic membrane stimulates the regeneration of the action potential in the postsynaptic neuron.

As secretory vesicles fuse with the cell membrane, the area of the cell membrane increases. Normal size is regained by the reuptake of membrane components through endocytosis. Regions bud in from the cell membrane and then fuse with internal membranes to effect recycling.

Sorting of Products by Chemical Receptors

Not all proteins synthesized on the ER are destined for export. Many, such as the hydrolases in lysosomes, remain inside the cell; others become anchored in the membrane of internal organelles or in the cell membrane. It is presumed that each protein has some type of marker that fits a specific location in the cell.

Proteins synthesized on free ribosomes have segments that bind to specific receptors on the outer membrane of mitochondria, chloroplasts, or peroxisomes, allowing these proteins to be taken up only by these organelles. In the case of proteins synthesized in the RER, both the hydrolases destined for lysosomes and the secretory proteins are found initially in the same portion of the ER lumen. Studies have shown that these can be distinguished on the basis of their carbohydrate residues. The carbohydrate residues of lysosomal enzymes become modified in the cis-Golgi by the addition of certain phosphate groups. This critical modification allows the enzymes to bind to specific receptors on the membrane of the Golgi, which then directs them into vesicles leading to a lysosome rather than a secretory vesicle. In the lysosomes, proton pumps create an acidic environment that causes the release of the lysosomal enzyme from the membrane-bound receptors. Much of this sorting activity is mediated by coated vesicles containing the same fibrous outer protein, clathrin, used in endocytosis. These sorting vesicles also contain associated smaller proteins.

Nucleus

The nucleus is the information centre of the cell and is surrounded by a nuclear membrane in all eukaryotic organisms. It is separated from the cytoplasm by the nuclear envelope, and it houses the double-stranded, spiral-shaped deoxyribonucleic acid (DNA) molecules, which contain the genetic information necessary for the cell to retain its unique character as it grows and divides.

The presence of a nucleus distinguishes the eukaryotic cells of multicellular organisms from the prokaryotic, one-celled organisms such as bacteria. In contrast to the higher organisms, prokaryotes do not have nuclei, so their DNA is maintained in the same compartment as their other cellular components.

The primary function of the nucleus is the expression of selected subsets of the genetic information encoded in the DNA double helix. Each subset of a DNA chain, called a gene, codes for the construction of a specific protein out of a chain of amino acids. Information in DNA is not decoded directly into proteins, however. First it is transcribed, or copied, into a range of messenger ribonucleic acid (mRNA) molecules, each of which encodes the information for one protein (or more than one protein in bacteria). The mRNA molecules are then transported through the nuclear envelope into the cytoplasm, where they are translated, serving as templates for the synthesis of specific proteins.

The nucleus must not only synthesize the mRNA for many thousands of proteins, but it must also regulate the amounts synthesized and supplied to the cytoplasm. Furthermore, the amounts of each type of mRNA supplied to the cytoplasm must be regulated differently in each type of cell. In addition to mRNA, the nucleus synthesizes and exports other classes of RNA involved in the mechanisms of protein synthesis.

Structural Organization of the Nucleus

DNA Packaging

The nucleus of the average human cell is only 6 micrometres (6×10^{-6} metre) in diameter, yet it contains about 1.8 metres of DNA. This is distributed among 46 chromosomes, each consisting of a single DNA molecule about 40 mm (1.5 inches) long. The extraordinary packaging problem this poses can be envisaged by a scale model enlarged a million times. On this scale a DNA molecule would be a thin string 2 mm thick, and the average chromosome would contain 40 km (25 miles) of DNA. With a diameter of only 6 metres, the nucleus would contain 1,800 km (1,118 miles) of DNA.

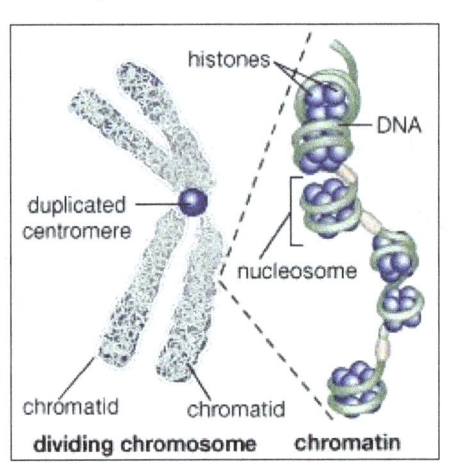

During the first stages of cell division, the recognizable double-stranded chromosome is formed by two tightly coiled DNA strands (chromatids) joined at a point called the centromere. During the middle stage of cell division, the centromere duplicates, and the chromatid pair separates. Following cell division, the separated chromatids uncoil; the loosely coiled DNA, wrapped around its associated proteins (histones) to form beaded structures called nucleosomes, is termed chromatin.

These contents must be organized in such a way that they can be copied into RNA accurately and selectively. DNA is not simply crammed or wound into the nucleus like a ball of string; rather, it is organized, by molecular interaction with specific nuclear proteins, into a precisely packaged structure. This combination of DNA with proteins creates a dense, compact fibre called chromatin. An extreme example of the ordered folding and compaction that chromatin can undergo is seen during cell division, when the chromatin of each chromosome condenses and is divided between two daughter cells.

Nucleosomes: The Subunits of Chromatin

The compaction of DNA is achieved by winding it around a series of small proteins called histones. Histones are composed of positively charged amino acids that bind tightly to and neutralize the negative charges of DNA. There are five classes of histone. Four of them, called H2A, H2B, H3, and H4, contribute two molecules each to form an octamer, an eight-part core around which two turns of DNA are wrapped. The resulting beadlike structure is called the nucleosome. The DNA enters and leaves a series of nucleosomes, linking them like beads along a string in lengths that vary between species of organism or even between different types of cell within a species. A string of nucleosomes is then coiled into a solenoid configuration by the fifth histone, called H1. One molecule of H1 binds to the site at which DNA enters and leaves each nucleosome, and a chain of H1 molecules coils the string of nucleosomes into the solenoid structure of the chromatin fibre.

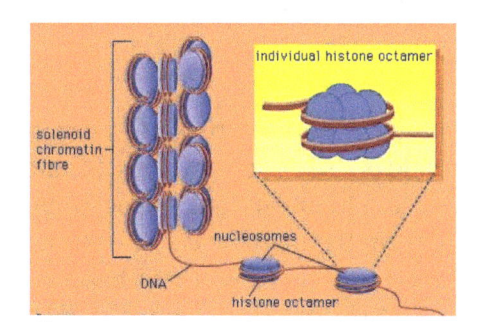

DNA wrapped around clusters of histone proteins to form nucleosomes, which can coil to form solenoids.

Nucleosomes not only neutralize the charges of DNA, but they have other consequences. First, they are an efficient means of packaging. DNA becomes compacted by a factor of six when wound into nucleosomes and by a factor of about 40 when the nucleosomes are coiled into a solenoid chromatin fibre. The winding into nucleosomes also allows some inactive DNA to be folded away in inaccessible conformations, a process that contributes to the selectivity of gene expression.

Organization of Chromatin Fibre

Several studies indicate that chromatin is organized into a series of large radial loops anchored to specific scaffold proteins. Each loop consists of a chain of nucleosomes and may be related to units of genetic organization. This radial arrangement of chromatin loops compacts DNA about a thousandfold. Further compaction is achieved by a coiling of the entire looped chromatin fibre into a dense structure called a chromatid, two of which form the chromosome. During cell division, this coiling produces a 10,000-fold compaction of DNA.

The Nuclear Envelope

The nuclear envelope is a double membrane composed of an outer and an inner phospholipid bilayer. The thin space between the two layers connects with the lumen of the rough endoplasmic reticulum (RER), and the outer layer is an extension of the outer face of the RER.

The inner surface of the nuclear envelope has a protein lining called the nuclear lamina, which binds to chromatin and other contents of the nucleus. The entire envelope is perforated by numerous nuclear pores. These transport routes are fully permeable to small molecules up to the size of the smallest proteins, but they form a selective barrier against movement of larger molecules. Each pore is surrounded by an elaborate protein structure called the nuclear pore complex, which selects molecules for entrance into the nucleus. Entering the nucleus through the pores are the nucleotide building blocks of DNA and RNA, as well as adenosine triphosphate, which provides the energy for synthesizing genetic material. Histones and other large proteins must also pass through the pores. These molecules have special amino acid sequences on their surface that signal admittance by the nuclear pore complexes. The complexes also regulate the export from the nucleus of RNA and subunits of ribosomes.

DNA in prokaryotes is also organized in loops and is bound to small proteins resembling histones, but these structures are not enclosed by a nuclear membrane.

Genetic Organization of the Nucleus

The Structure of DNA

Several features are common to the genetic structure of most organisms. First is the double-stranded DNA. Each strand of this molecule is a series of nucleotides, and each nucleotide is composed of a sugar-phosphate compound attached to one of four nitrogen-containing bases. The sugar-phosphate compounds link together to form the backbone of the strand. Each of the bases strung along the backbone is chemically attracted to a corresponding base on the parallel strand of the DNA molecule. This base pairing joins the two strands of the molecule much as rungs join the two sides of a ladder, and the chemical bonding of the base pairs twists the doubled strands into a spiral, or helical, shape.

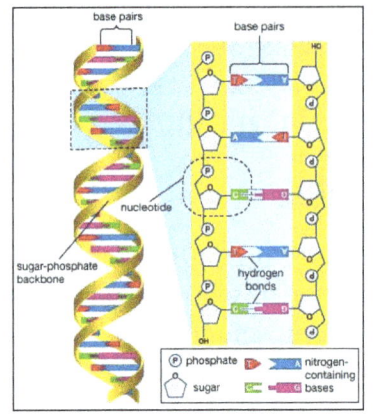

DNA molecule

The four nucleotide bases are adenine, cytosine, guanine, and thymine. DNA is composed of millions of these bases strung in an apparently limitless variety of sequences. It is in the sequence of bases that the genetic information is contained, each sequence determining the sequence of amino acids to be connected into proteins. A nucleotide sequence sufficient to encode one protein is called a gene. Genes are interspersed along the DNA molecule with other sequences that do not encode proteins. Some of these so-called untranslated regions regulate the activity of the adjacent genes, for example, by marking the points at which enzymes begin and cease transcribing DNA into RNA.

Rearrangement and Modification of DNA

Rearrangements and modifications of the nucleotide sequences in DNA are exceptions to the rules of genetic expression and sometimes cause significant changes in the structure and function of cells. Different cells of the body owe their specialized structures and functions to different genes. This does not mean that the set of genetic information varies among the cells of the body. Indeed, for each cell the entire DNA content of the chromosomes is usually duplicated exactly from generation to generation, and, in general, the genetic content and arrangement is strikingly similar among different cell types of the same organism. As a result, the differentiation of cells can occur without the loss or irreversible inactivation of unnecessary genes, an observation that is reinforced by the presence of specific genes in a range of adult tissues. For example, normal copies of the genes encoding hemoglobin are present in the same numbers in red blood cells, which make hemoglobin, as in a range of other types of cells, which do not.

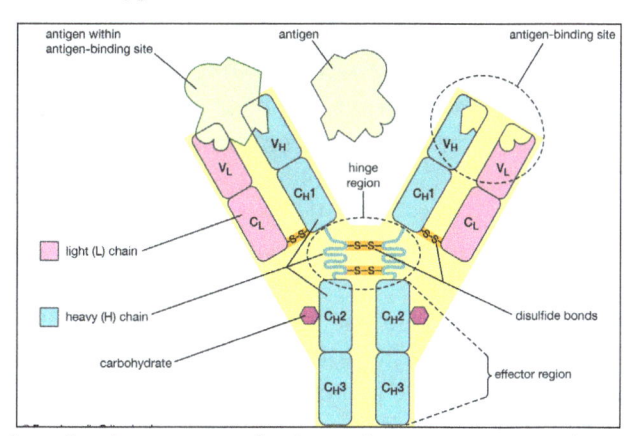

The structure of an antibody molecule represents the dramatic rearrangements of DNA that occur in the immune systems of mammals. Each antibody contains a light chain and a heavy chain that are encoded by different segments of DNA. These segments are subject to considerable variation and are thus able to produce many different antibodies.

Despite the general uniformity of genetic content in all the cells of an organism, studies have shown a few clear examples in some organisms of programmed, reversible change in the DNA of developing tissues. One of the most dramatic rearrangements of DNA occurs in the immune systems of mammals. The body's defense against invasion by foreign organisms involves the synthesis of a vast range of antibodies by lymphocytes (a type of white blood cell). Antibodies are proteins that bind to specific invading molecules or organisms and either inactivate them or signal their destruction. The binding sites on each antibody molecule are formed by one light and one heavy amino acid chain, which are encoded by different segments of the DNA in the lymphocyte nucleus. These DNA segments undergo considerable rearrangements, resulting in the synthesis of a great variety of antibodies. Some invasive organisms, such as trypanosome parasites, which cause sleeping sickness, go to great lengths to rearrange their own DNA to evade the versatility of their hosts' antibody production. The parasites are covered by a thick coat of glycoprotein (a protein with sugars attached). Given time, host organisms can overcome infection by producing antibodies to the parasites' glycoprotein coat, but this reaction is anticipated and evaded by the selective rearrangement of the trypanosomes' DNA encoding the glycoprotein, thus constantly changing the surface presented to the hosts' immune system.

Careful comparisons of gene structure have also revealed epigenetic modifications, heritable changes that occur on the sugar-phosphate side of bases in the DNA and thus do not cause rearrangements in the DNA sequence itself. An example of an epigenetic modification involves the addition of a methyl group to cytosine bases. This appears to cause the inactivation of genes that do not need to be expressed in a particular type of cell. An important feature of the methylation of cytosine lies in its ability to be copied, so that methyl groups in a dividing cell's DNA will result in methyl groups in the same positions in the DNA of both daughter cells.

Genetic Expression through RNA

The transcription of the genetic code from DNA to RNA, and the translation of that code from RNA into protein, exerts the greatest influence on the modulation of genetic information. The process of genetic expression takes place over several stages, and at each stage is the potential for further differentiation of cell types.

As explained above, genetic information is encoded in the sequences of the four nucleotide bases making up a DNA molecule. One of the two DNA strands is transcribed exactly into messenger RNA (mRNA), with the exception that the thymine base of DNA is replaced by uracil. RNA also contains a slightly different sugar component (ribose) from that of DNA (deoxyribose) in its connecting sugar-phosphate chain. Unlike DNA, which is stable throughout the cell's life and of which individual strands are even passed on to many cell generations, RNA is unstable. It is continuously broken down and replaced, enabling the cell to change its patterns of protein synthesis.

Apart from mRNA, which encodes proteins, other classes of RNA are made by the nucleus. These include ribosomal RNA (rRNA), which forms part of the ribosomes and is exported to the cytoplasm to help translate the information in mRNA into proteins. Ribosomal RNA is synthesized in a specialized region of the nucleus called the nucleolus, which appears as a dense area within the nucleus and contains the genes that encode rRNA. This is also the site of assembly of ribosome subunits from rRNA and ribosomal proteins. Ribosomal proteins are synthesized in the cytoplasm and transported to the nucleus for subassembly in the nucleolus. The subunits are then returned to the cytoplasm for final assembly. Another class of RNA synthesized in the nucleus is transfer RNA (tRNA), which serves as an adaptor, matching individual amino acids to the nucleotide triplets of mRNA during protein synthesis.

RNA Synthesis

The synthesis of RNA is performed by enzymes called RNA polymerases. In higher organisms there are three main RNA polymerases, designated I, II, and III (or sometimes A, B, and C). Each is a complex protein consisting of many subunits. RNA polymerase I synthesizes three of the four types of rRNA (called 18S, 28S, and 5.8S RNA); therefore it is active in the nucleolus, where the genes encoding these rRNA molecules reside. RNA polymerase II synthesizes mRNA, though its initial products are not mature RNA but larger precursors, called heterogeneous nuclear RNA, which are completed later. The products of RNA polymerase III include tRNA and the fourth RNA component of the ribosome, called 5S RNA.

All three polymerases start RNA synthesis at specific sites on DNA and proceed along the molecule, linking selected nucleotides sequentially until they come to the end of the gene and terminate the growing chain of RNA. Energy for RNA synthesis comes from high-energy phosphate linkages contained in the nucleotide precursors of RNA. Each unit of the final RNA product is essentially a sugar, a base, and one phosphate, but the building material consists of a sugar, a base, and three phosphates. During synthesis two phosphates are cleaved and discarded for each nucleotide that is incorporated into RNA. The energy released from the phosphate bonds is used to link the nucleotides. The crucial feature of RNA synthesis is that the sequence of nucleotides joined into a growing RNA chain is specified by the sequence of nucleotides in the DNA template: each adenine in DNA specifies uracil in RNA, each cytosine specifies guanine, each guanine specifies cytosine, and each thymine in DNA specifies adenine. In this way the information encoded in each gene is transcribed into RNA for translation by the protein-synthesizing machinery of the cytoplasm.

In addition to specifying the sequence of amino acids to be polymerized into proteins, the nucleotide sequence of DNA contains supplementary information. For example, short sequences of nucleotides determine the initiation site for each RNA polymerase, specifying where and when RNA synthesis should occur. In the case of RNA polymerases I and II, the sequences specifying initiation sites lie just ahead of the genes. In contrast, the equivalent information for RNA polymerase III lies within the gene—that is, within the region of DNA to be copied into RNA. The initiation site on a segment of DNA is called a promoter. The promoters of different genes have some nucleotide sequences in common, but they differ in others. The differences in sequence are recognized by specific proteins called transcription factors, which are necessary for the expression of particular types of genes. The specificity of transcription factors contributes to differences in the gene expression of different types of cells.

Processing of mRNA

During and after synthesis, mRNA precursors undergo a complex series of changes before the mature molecules are released from the nucleus. First, a modified nucleotide is added to the start of the RNA molecule by a reaction called capping. This cap later binds to a ribosome in the cytoplasm. The synthesis of mRNA is not terminated simply by the RNA polymerase's detachment from DNA, but by chemical cleavage of the RNA chain. Many (but not all) types of mRNA have a simple polymer of adenosine residues added to their cleaved ends.

In addition to these modifications of the termini, startling discoveries in 1977 revealed that portions of newly synthesized RNA molecules are cut out and discarded. In many genes, the regions coding for proteins are interrupted by intervening sequences of nucleotides called introns. These introns must be excised from the RNA copy before it can be released from the nucleus as a functional mRNA. The number and size of introns within a gene vary greatly, from no introns at all to more than 50. The sum of the lengths of these intervening sequences is sometimes longer than the sum of the regions coding for proteins.

The removal of introns, called RNA splicing, appears to be mediated by small nuclear ribonucleoprotein particles (snRNP's). These particles have RNA sequences that are complementary to the junctions between introns and adjacent coding regions. By binding to the junction ends, an snRNP twists the intron into a loop. It then excises the loop and splices the coding regions.

Regulation of Genetic Expression

Although all the cell nuclei of an organism generally carry the same genes, there are conspicuous differences between the specialized cell types of the body. The source of these differences lies not so much in the occasional modification of DNA, as outlined above, but in the selective expression of DNA through RNA; in particular, it can be traced to processes regulating the amounts and activities of mRNA both during and after its synthesis in the nucleus.

Regulation of RNA Synthesis

The first level of regulation is mediated by variations in chromatin structure. In order to be transcribed, a gene must be assembled into a structurally distinct form of active chromatin. A second level of regulation is achieved by varying the frequency with which a gene in the active conformation is transcribed into RNA by an RNA polymerase. There is evidence for regulation of RNA synthesis at both these levels—for example, in response to hormone induction. At both levels, protein factors are believed to perform the regulation—for example, by binding to special promoter DNA regions flanking the transcribed gene.

Regulation of RNA after Synthesis

After synthesis, RNA molecules undergo selective processing, which results in the export of only a subpopulation of RNA molecules to the cytoplasm. Furthermore, the stability in the cytoplasm of a particular type of mRNA can be regulated. For example, the hormone prolactin increases synthesis of milk proteins in tissue by causing a twofold rise in the rate of mRNA synthesis; but it also causes a 17-fold rise in mRNA lifetime, so that in this case the main cause of increased protein synthesis is the prolonged availability of mRNA. Conversely, there is evidence for selective destabilization of some mRNA—such as histone mRNA, which is rapidly broken down when DNA replication is interrupted. Finally, there are many examples of selective regulation of the translation of mRNA into protein.

Mitochondrion and Chloroplast

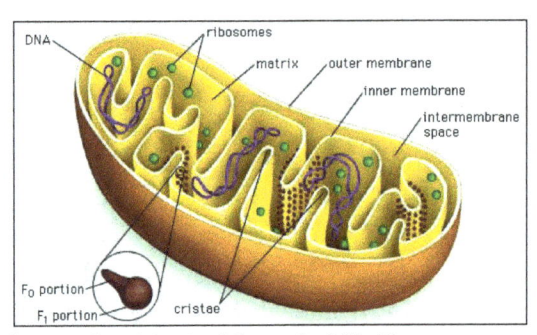

The internal membrane of a mitochondrion is elaborately folded into structures known as cristae. Cristae increase the surface area of the inner membrane, which houses the components of the electron-transport chain. Proteins known as F_1F_0ATPases that produce the majority of ATP used by cells are found throughout the cristae.

Mitochondria and chloroplasts are the powerhouses of the cell. Mitochondria appear in both plant and animal cells as elongated cylindrical bodies, roughly one micrometre in length and closely packed in regions actively using metabolic energy. Mitochondria oxidize the products of cytoplasmic metabolism to generate adenosine triphosphate (ATP), the energy currency of the cell. Chloroplasts are the photosynthetic organelles in plants and some algae. They trap light energy and convert it partly into ATP but mainly into certain chemically reduced molecules that, together with ATP, are used in the first steps of carbohydrate production. Mitochondria and chloroplasts share a certain structural resemblance, and both have a somewhat independent existence within the cell, synthesizing some proteins from instructions supplied by their own DNA.

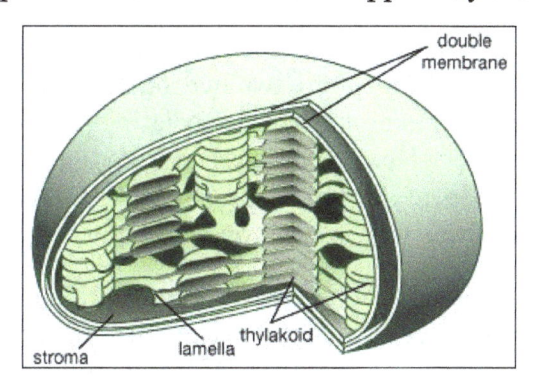

Internal structures of the chloroplast The interior contains flattened sacs of photosynthetic membranes (thylakoids) formed by the invagination and fusion of the inner membrane. Thylakoids are usually arranged in stacks (grana) and contain the photosynthetic pigment (chlorophyll). The grana are connected to other stacks by simple membranes (lamellae) within the stroma, the fluid proteinaceous portion containing the enzymes essential for the photosynthetic dark reaction, or Calvin cycle.

Mitochondrial and Chloroplastic Structure

Both organelles are bounded by an external membrane that serves as a barrier by blocking the passage of cytoplasmic proteins into the organelle. An inner membrane provides an additional barrier that is impermeable even to small ions such as protons. The membranes of both organelles have a lipid bilayer construction. Located between the inner and outer membranes is the intermembrane space.

In mitochondria the inner membrane is elaborately folded into structures called cristae that dramatically increase the surface area of the membrane. In contrast, the inner membrane of chloroplasts is relatively smooth. However, within this membrane is yet another series of folded membranes that form a set of flattened, disklike sacs called thylakoids. The space enclosed by the inner membrane is called the matrix in mitochondria and the stroma in chloroplasts. Both spaces are filled with a fluid containing a rich mixture of metabolic products, enzymes, and ions. Enclosed by the thylakoid membrane of the chloroplast is the thylakoid space. The extraordinary chemical capabilities of the two organelles lie in the cristae and the thylakoids. Both membranes are studded with enzymatic proteins either traversing the bilayer or dissolved within the bilayer. These proteins contribute to the production of energy by transporting material across the membranes and by serving as electron carriers in important oxidation-reduction reactions.

Metabolic Functions

Crucial to the function of mitochondria and chloroplasts is the chemistry of the oxidation-reduction, or redox, reaction. This controlled burning of material comprises the transfer of electrons from one compound, called the donor, to another, called the acceptor. All compounds taking part in redox reactions are ranked in a descending scale according to their ability to act as electron donors. Those higher in the scale donate electrons to their fellows lower down, which have a lesser tendency to donate, but a correspondingly greater tendency to accept, electrons. Each acceptor in turn donates electrons to the next compound down the scale, forming a donor-acceptor chain extending from the greatest donating ability to the least.

At the top of the scale is hydrogen, the most abundant element in the universe. The nucleus of a hydrogen atom is composed of one positively charged proton; around the nucleus revolves one negatively charged electron. In the atmosphere two hydrogen atoms join to form a hydrogen molecule (H_2). In solution the two atoms pull apart, dissociating into their constituent protons and electrons. In the redox reaction the electrons are passed from one reactant to another. The donation of electrons is called oxidation, and the acceptance is called reduction—hence the descriptive term oxidation-reduction, indicating that one action never takes place without the other.

A hydrogen atom has a great tendency to transfer an electron to an acceptor. An oxygen atom, in contrast, has a great tendency to accept an electron. The burning of hydrogen by oxygen is, chemically, the transfer of an electron from each of two hydrogen atoms to oxygen, so that hydrogen is oxidized and oxygen reduced. The reaction is extremely exergonic; i.e., it liberates much free energy as heat. This is the reaction that takes place within mitochondria but is so controlled that the heat is liberated not at once but in a series of steps. The free energy, harnessed by the organelle, is coupled to the synthesis of ATP from adenosine diphosphate (ADP) and inorganic phosphate (Pi).

An analogy can be drawn between this controlled reaction and the flow of river water down a lock system. Without the locks, water flow would be rapid and uncontrolled, and no ship could safely ply the river. The locks force water to flow in small controlled steps conducive to safe navigation. But there is more to a lock system than this. The flow of water down the locks can also be harnessed to raise a ship from a lower to a higher level, with the water rather than the ship expending the energy. In mitochondria the burning of hydrogen is broken into a series of small indirect steps following the flow of electrons along a chain of donor-acceptors. Energy is funneled into the chemical bonding of ADP and Pi, raising the free energy of these two compounds to the high level of ATP.

The Mitochondrion

Formation of the Electron Donors NADH and FADH2

Through a series of metabolic reactions carried out in the matrix, the mitochondrion converts products of the cell's initial metabolism of fats, amino acids, and sugars into the compound acetyl coenzyme A. The acetate portion of this compound is then oxidized in a chain reaction called the tricarboxylic acid cycle. At the end of this cycle the carbon atoms yield carbon dioxide and the hydrogen atoms are transferred to the cell's most important hydrogen acceptors, the coenzymes nicotinamide adenine dinucleotide (NAD+) and flavin adenine dinucleotide (FAD), yielding NADH and FADH2. It is the subsequent oxidation of these hydrogen acceptors that leads eventually to the production of ATP.

NADH and FADH2 are compounds of high electron-donating capacity. Were they to transfer their electrons directly to oxygen, the resulting combustion would release a lethal burst of heat energy. Instead, the energy is released in a series of electron donor-acceptor reactions carried out within the cristae of the mitochondrion by a number of proteins and coenzymes that make up the electron-transport, or respiratory, chain.

The Electron-transport Chain

The proteins of this chain are embedded in the cristae membrane, actually traversing the lipid bilayer and protruding from the inner and outer surfaces. The coenzymes are dissolved in the lipid and diffuse through the membrane or across its surface. The proteins are arranged in three large complexes, each composed of a number of polypeptide chains. Each complex is, to continue the hydraulic analogy, a lock in the waterfall of the electron flow and the site at which energy from the overall redox reaction is tapped. The first complex, NADH dehydrogenase, accepts a pair of electrons from the primary electron donor NADH and is reduced in the process. It in turn donates these electrons to the coenzyme ubiquinone, a lipid-soluble molecule composed of a substituted benzene ring attached to a hydrocarbon tail. Ubiquinone, diffusing through the lipid of the cristae membrane, reaches the second large complex of the electron-transport chain, the b-c2 complex, which accepts the electrons, oxidizing ubiquinone and being itself reduced. (This complex can also accept electrons from the second primary electron donor, FADH2, a molecule below NADH in the electron-donating scale.) The b-c2 complex transfers the pair of electrons to cytochrome c, a small protein situated on the outer surface of the cristae membrane. From cytochrome c, electrons pass (four at a time) to the third large complex, cytochrome oxidase, which, in the final step of the chain, transfers the four electrons to two oxygen atoms and two protons, generating two water molecules.

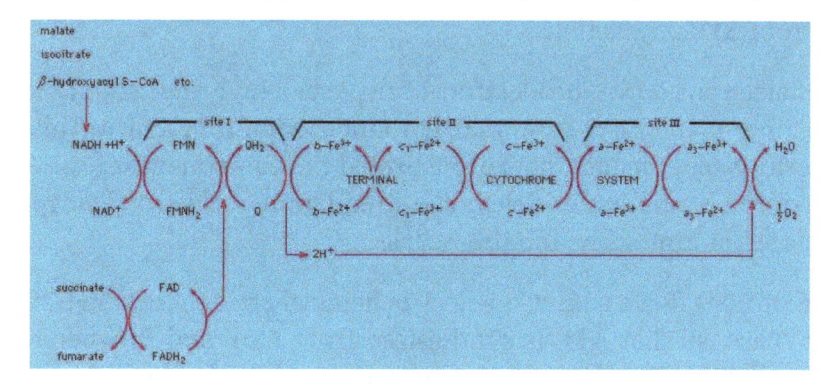

The electron-transport chain embedded in the inner membrane of a mitochondrion is made up of a series of electron donors and electron acceptors. The transport of electrons begins with the acceptance of electrons by NADH dehydrogenase from NADH. The electrons are then passed to ubiquinone (coenzyme Q; site I), which carries them to the b-c2 complex. The electrons are then transferred to cytochrome c (site II), to cytochrome oxidase (site III), and finally to oxygen.

This transfer of electrons, from member to member of the electron-transport chain, provides energy for the synthesis of ATP through an indirect route. At the beginning of the electron-transport chain, NADH and $FADH_2$ split hydrogen atoms into protons and electrons, transferring the electrons to the next protein complex and releasing the protons into the mitochondrial matrix. When each protein complex in turn transfers the electrons down the chain, it uses the energy released in this process to pump protons across the inner membrane into the intermembrane space. This transport of positively charged protons into the intermembrane space, opposite the negatively charged electrons in the matrix, creates an electrical potential that tends to draw the protons back across the membrane. A high concentration of protons outside the membrane also creates the conditions for their diffusion back into the matrix. However, as explained above, the inner membrane is extremely impermeable to protons. In order for the protons to flow back down the electrochemical gradient, they must traverse the membrane through transport molecules similar to the protein complexes of the electron-transfer chain. These molecules are the so-called F_1F_0ATPase, a complex protein that, transporting protons back into the matrix, uses the energy released to synthesize ATP. The protons then join the electrons and oxygen atoms to form water.

This complex chain of events, the basis of the cell's ability to derive ATP from metabolic oxidation, was conceived in its entirety by the British biochemist Peter Mitchell in 1961. The years following the announcement of his chemiosmotic theory saw its ample substantiation and revealed its profound implications for cell biology.

The Chemiosmotic Theory

The four postulates of the chemiosmotic theory, including examples of their experimental substantiation, are as follows:

1. The inner mitochondrial membrane is impermeable to protons, hydroxide ions, and other cations and anions. This postulate was validated when it was shown that substances allowing protons to flow readily across mitochondrial membranes uncouple oxidative electron transport from ATP production.

2. Transfer of electrons down the electron-transport chain brings about pumping of protons across the inner membrane, from matrix to intermembrane space. This was demonstrated in laboratory experiments that reconstituted the components of the electron-transport chain in artificial membrane vesicles. The stimulation of electron transport caused a measurable buildup of protons within the vesicle.

3. The flow of protons down a built-up electrochemical gradient occurs through a proton-dependent ATPase, so that ATP is synthesized from ADP and Pi whenever protons move through the enzyme. This hypothesis was confirmed by the discovery of what came to be known as the F_1F_0ATPase. Shaped like a knob attached to the membrane by a narrow stalk, F_1F_0ATPase covers the inner surface of the cristae. Its stalk (the F_0 portion) penetrates the lipid bilayer of the inner membrane and is capable of catalyzing the transport of protons. The knob (the F_1 portion) is capable of synthesizing as well as splitting, or hydrolyzing, ATP. F_1F_0ATPase is therefore reversible, either using the energy of proton diffusion to combine ADP and Pi or using the energy of ATP hydrolysis to pump protons out of the matrix.

4. The inner membrane of the mitochondrion possesses a complement of proteins that brings about the transport of essential metabolites. Numerous carrier systems have been demonstrated to transport into the mitochondrion the products of metabolism that are transformed into substrates for the electron-transport chain. Best known is the ATP-ADP exchange carrier of the inner membrane. Neither ATP nor ADP, being large charged molecules, can cross the membrane unaided, but ADP must enter and ATP must leave the mitochondrial matrix in order for ATP synthesis to continue. A single protein conducts the

counter-transport of ATP against ADP, the energy released by the flow of ATP down its concentration gradient being coupled to the pumping of ADP up its gradient and into the mitochondrion.

The Chloroplast

Trapping of Light

Light travels as packets of energy known as photons and is absorbed in this form by light-absorbing chlorophyll molecules embedded in the thylakoid membrane of the chloroplast. The chlorophyll molecules are grouped into antenna complexes, clusters of several hundred molecules that are anchored onto the thylakoid membrane by special proteins. Within each antenna complex is a specialized set of proteins and chlorophyll molecules that form a reaction centre. Photons absorbed by the other chlorophylls of the antenna are funneled into the reaction centre. The energy of the photon is absorbed by an electron of the reaction centre molecule in sufficient quantity to enable its acceptance by a nearby coenzyme, which cannot accept electrons at low energy levels. This coenzyme has a high electron-donor capability; it initiates the transfer of the electron down an electron-transport chain similar to that of the mitochondrion. Meanwhile, the loss of the negatively charged electron leaves a positively charged "hole" in the reaction centre chlorophyll molecule. This hole is filled by the enzymatic splitting of water into molecular oxygen, protons, and electrons and the transfer of an electron to the chlorophyll. The oxygen is released by the chloroplast, making its way out of the plant and into the atmosphere. The protons, in a process similar to that in the mitochondrion, are pumped through the thylakoid membrane and into the thylakoid space. Their facilitated diffusion back into the stroma through proteins embedded in the membrane powers the synthesis of ATP. This part of the photosynthetic process is called photosystem II.

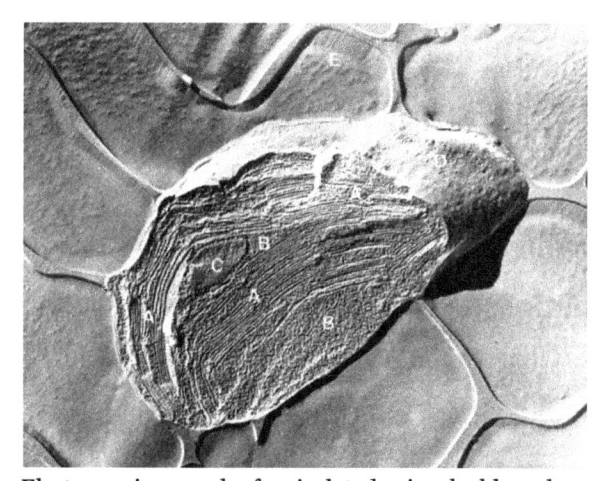

Electron micrograph of an isolated spinach chloroplast.

At the end of the electron-transport chain in the thylakoid membrane is another reaction centre molecule. The electron is again energized by photons and then transported down another chain, which makes up photosystem I. This system uses the energy released in electron transfer to join a proton to nicotinamide adenine dinucleotide phosphate ($NADP^+$), a phosphorylated derivative of NAD^+, forming NADPH. NADPH is a high-energy electron donor that, with ATP, fuels the conversion of carbon dioxide into the carbohydrate foods of the plant cell.

The reaction of photosynthesis

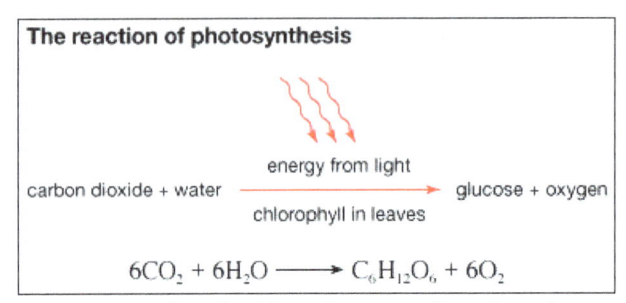

carbon dioxide + water $\xrightarrow[\text{chlorophyll in leaves}]{\text{energy from light}}$ glucose + oxygen

$6CO_2 + 6H_2O \longrightarrow C_6H_{12}O_6 + 6O_2$

In photosynthesis, plants consume carbon dioxide and water and produce glucose and oxygen. Energy for this process is provided by light, which is absorbed by pigments, primarily chlorophyll. Chlorophyll is the pigment that gives plants their green colour.

Fixation of Carbon Dioxide

NADPH remains within the stroma of the chloroplast for use in the fixation of carbon dioxide (CO_2) during the Calvin cycle. In a complex cycle of chemical reactions, CO_2 is bound to a five-carbon ribulose biphosphate compound. The resulting six-carbon intermediate is then split into three-carbon phosphoglycerate. With energy supplied by the breakdown of NADPH and ATP, this compound is eventually formed into glyceraldehyde 3-phosphate, an important sugar intermediate of metabolism. One glyceraldehyde molecule is exported from the chloroplast, for further conversion in the cytoplasm, for every five that undergo an ATP-powered re-formation into the five-carbon ribulose biphosphate. In this way three molecules of CO_2 yield one molecule of glyceraldehyde 3-phosphate, while the entire fixation cycle hydrolyzes nine molecules of ATP and oxidizes six molecules of NADPH.

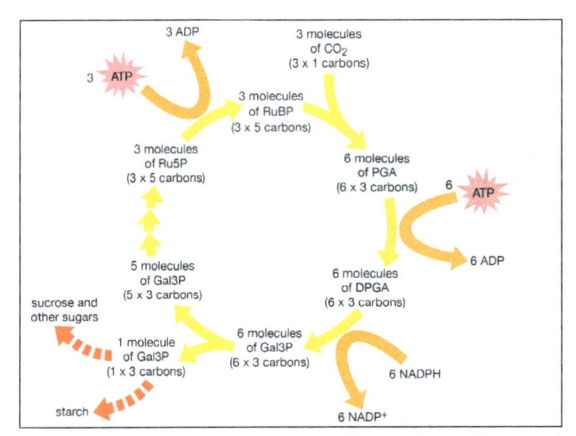

Pathway of carbon dioxide fixation and reduction in photosynthesis, the Calvin cycle. The diagram represents one complete turn of the cycle, with the net production of one molecule of glyceraldehyde-3-phosphate (Gal3P). This three-carbon sugar phosphate usually is converted to either sucrose or starch.

Evolutionary Origins

The Mitochondrion and Chloroplast as Independent Entities

In addition to their remarkable metabolic capabilities, both mitochondria and chloroplasts synthesize on their own a number of proteins and lipids necessary for their structure and activity. Not only do they contain the machinery necessary for this, but they also possess the genetic material to direct it. DNA within these organelles has a circular structure reminiscent of prokaryotic, not eukaryotic, DNA. Also as in prokaryotes, the DNA is not associated with histones. Along with the DNA are protein-synthesizing ribosomes, of prokaryotic rather than eukaryotic size.

Only a small portion of the mitochondrion's total number of proteins is synthesized within the organelle. Numerous proteins are encoded and made in the cytoplasm specifically for export into the mitochondrion. The mitochondrial DNA itself encodes only 13 different proteins. The proteins that contain subunits synthesized within the mitochondrion often also possess subunits synthesized in the cytoplasm. Mitochondrial and chloroplastic proteins synthesized in the cytoplasm have to enter the organelle by a complex process, crossing both the outer and the inner membranes. These proteins contain specific arrangements of amino acids known as leader sequences that are recognized by receptors on the outer membranes of the organelles. The proteins are then guided through membrane channels in an energy-requiring process.

The Endosymbiont Hypothesis

Mitochondria and chloroplasts are self-dividing; they contain their own DNA and protein-synthesizing machinery, similar to that of prokaryotes. Chloroplasts produce ATP and trap photons by mechanisms that are complex and yet similar to those of certain prokaryotes. These phenomena have led to the theory that the two organelles are direct descendants of prokaryotes that entered primitive nucleated cells. Among billions of such events, a few could have led to the development of stable, symbiotic associations between nucleated hosts and prokaryotic parasites. The hosts would provide the parasites with a stable osmotic environment and easy access to nutrients, and the parasites would repay the hosts by providing an oxidative ATP-producing system or a photosynthetic energy-producing reaction.

The Cytoskeleton

The cytoskeleton is the name given to the fibrous network formed by different types of long protein filaments present throughout the cytoplasm of eukaryotic cells (cells containing a nucleus). The filaments of the cytoskeleton create a scaffold, or framework, that organizes other cell constituents and maintains the shape of the cell. In addition, some filaments cause coherent movements, both of the cell itself and of its internal organelles. Prokaryotic (nonnucleated) cells, which are generally much smaller than eukaryotic cells, contain a unique related set of filaments but, with few exceptions, do not possess true cytoskeletons. Their shapes and the shapes of certain eukaryotes, primarily yeast and other fungi, are determined by the rigid cell wall on the outside of the cell.

Four major types of cytoskeletal filaments are commonly recognized: actin filaments, microtubules, intermediate filaments, and septins. Actin filaments and microtubules are dynamic structures that continuously assemble and disassemble in most cells. Intermediate filaments are stabler and seem to be involved mainly in reinforcing cell structures, especially the position of the nucleus and the junctions that connect cells. Septins are involved in cell division and have been implicated in other cell functions. A wide variety of accessory proteins works in concert with each type of filament, linking filaments to one another and to the cell membrane and helping to form the networks that endow the cytoskeleton with its unique functions. Many of these accessory proteins have been characterized, revealing a rich diversity in the structure and function of the cytoskeleton.

Actin Filaments

Actin is a globular protein that polymerizes (joins together many small molecules) to form long filaments. Because each actin subunit faces in the same direction, the actin filament is polar, with different ends, termed "barbed" and "pointed." An abundant protein in nearly all eukaryotic cells, actin has been extensively studied in muscle cells. In muscle cells, the actin filaments are organized into regular arrays that are complementary with a set of thicker filaments formed from a second protein called myosin. These two proteins create the force responsible for muscle contraction. When the signal to contract is sent along a nerve to the muscle, the actin and myosin are activated. Myosin works as a motor, hydrolyzing adenosine triphosphate (ATP) to release energy in such a way that a myosin filament moves along an actin filament, causing the two filaments to slide past each other. The thin actin filaments and the thick myosin filaments are organized in a structure called the sarcomere, which shortens as the filaments slide over one another. Skeletal muscles are composed of bundles of many long muscle cells; when the sarcomeres contract, each of these giant muscle cells shortens, and the overall effect is the contraction of the entire muscle. Although the stimulation pathways differ, heart muscle and smooth muscle (found in many internal organs and blood vessels) contract by a similar sliding filament mechanism.

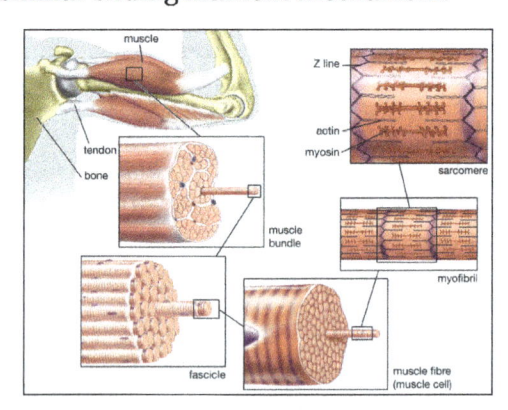

The structure of striated muscle striated muscle tissue, such as the tissue of the human biceps muscle, consists of long, fine fibres, each of which is in effect a bundle of finer myofibrils. Within each myofibril are filaments of the proteins myosin and actin; these filaments slide past one another as the muscle contracts and expands. On each myofibril, regularly occurring dark bands, called Z lines, can be seen where actin and myosin filaments overlap. The region between two Z lines is called a sarcomere; sarcomeres can be considered the primary structural and functional unit of muscle tissue.

Actin is also present in non-muscle cells, where it forms a meshwork of filaments responsible for many types of cellular movement. The meshwork consists of actin filaments that are attached to the cell membrane and to each other. The length of the filaments and the architecture of their attachments determine the shape and consistency of a cell. A large number of accessory proteins bind to actin, controlling the number, length, position, and attachments of the actin filaments. Different cells and tissues contain different accessory proteins, which accounts for the different shapes and movements of different cells. For example, in some cells, actin filaments are bundled by accessory proteins, and the bundle is attached to the cell membrane to form microvilli, stable protrusions that resemble tiny bristles. Microvilli on the surface of epithelial cells such as those lining the intestine increase the cell's surface area and thus facilitate the absorption of ingested food and water molecules. Other types of microvilli are involved in the detection of sound in the ear, where their movement, caused by sound waves, sends an electrical signal to the brain.

Many actin filaments in non-muscle cells have only a transient existence, polymerizing and depolymerizing in controlled ways that create movement. For example, many cells continually send out and retract tiny "filopodia," long needlelike projections of the cell membrane that are thought to enable cells to probe their environment and decide which direction to go. Like microvilli, filopodia are formed when actin filaments push out the membrane, but, because these actin filaments are less stable, filopodia have only a brief existence. Another actin structure only transiently associated with the cell membrane is the contractile ring, which is composed of actin filaments running around the circumference of the cell during cell division. As its name implies, this ring pulls in the cell membrane by a myosin-dependent process, thereby pinching the cell in half.

Microtubules

Microtubules are long filaments formed from 13 to 15 protofilament strands of a globular subunit called tubulin, with the strands arranged in the form of a hollow cylinder. Like actin filaments, microtubules are polar, having "plus" and "minus" ends. Most microtubule plus ends are constantly growing and shrinking, by respectively adding and losing subunits at their ends. Stable microtubules are found in cilia and flagella. Cilia are hairlike structures found on the surface of certain types of epithelial cells, where they beat in unison to move fluid and particles over the cell surface. Cilia are closely related in structure to flagella. Flagella such as those found on sperm cells produce a helical wavelike motion that enables a cell to propel itself rapidly through fluids. In cilia and flagella a set of microtubules is connected in a regular array by numerous accessory proteins that act as links and spokes in the assembly. Movement of the cilia or flagella occurs when adjacent microtubules slide past one another, bending the structures. This motion is caused by the motor protein dynein, which uses the energy of ATP hydrolysis to move along the microtubules, in a manner resembling the movement of myosin along actin filaments.

In most cells, microtubules grow outward, from the cell centre to the cell membrane, from a special region of the cytoplasm near the nuclear envelope called the centrosome. The minus ends of these microtubules are embedded in the centrosome, while the plus ends terminate near the cell membrane. The plus ends grow and shrink rapidly, a process known as dynamic instability. At the start of cell division, the centrosome replicates and divides in two. The two centrosomes separate and move to opposite sides of the nuclear envelope, where each nucleates a starlike array of microtubules, forming the mitotic spindle. The mitotic spindle partitions the duplicated chromosomes into the two daughter cells during mitosis.

Microtubules often serve as tracks for the transport of membrane vesicles in the cell, carried by the motor proteins kinesin and dynein. Kinesins generally move toward the plus end of the microtubule, and dyneins move toward the minus end. Microtubule-based vesicle transport occurs in nearly all cells, but it is especially prominent in the long thin processes of neurons, carrying essential components to and from the synapses at the ends of the processes.

Intermediate Filaments

Intermediate filaments are so named because they are thicker than actin filaments and thinner than microtubules or muscle myosin filaments. The subunits of intermediate filaments are elongated, not globular, and are associated in an antipolar manner. As a result, the overall filament has no polarity, and therefore no motor proteins move along intermediate filaments. Intermediate filaments are found only in complex multicellular organisms. They are encoded by a large number of different genes and can be grouped into families based on their amino acid sequences. Cells in different tissues of the body express one or another of these genes at different times. One cell can even change which type of intermediate filament protein is expressed over its lifetime. Most likely, the different forms of intermediate filaments have subtle but critical differences in their functional characteristics, helping to define the function of the cell. In general, intermediate filaments serve as structural elements, helping cells maintain their shape and integrity. For example, keratin filaments, the intermediate filaments of epithelial cells, which line surfaces of the body, give strength to the cell sheet that covers the surface. Mutations in keratin genes can result in blisters when the epithelial cell sheet is weak and prone to rupture. Keratin mutations can also cause deformations in the hair, nails, and corneas. Another example of a family of intermediate filaments is the lamin family, which comprises the nuclear lamina, a fibrous shell that underlies and supports the nuclear membrane.

Cell Division and Growth

In unicellular organisms, cell division is the means of reproduction; in multicellular organisms, it is the means of tissue growth and maintenance. Survival of the eukaryotes depends upon interactions between many cell types, and it is essential that a balanced distribution of types be maintained. This is achieved by the highly regulated process of cell proliferation. The growth and division of different cell populations are regulated in different ways, but the basic mechanisms are similar throughout multicellular organisms.

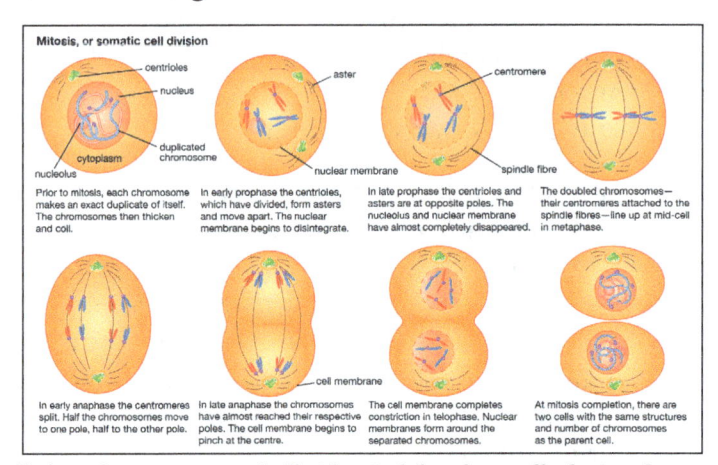

Mitosis: One cell gives rise to two genetically identical daughter cells during the process of mitosis.

Most tissues of the body grow by increasing their cell number, but this growth is highly regulated to maintain a balance between different tissues. In adults most cell division is involved in tissue renewal rather than growth, many types of cells undergoing continuous replacement. Skin cells, for example, are constantly being sloughed off and replaced; in this case, the mature differentiated cells do not divide, but their population is renewed by division of immature stem cells. In certain other cells, such as those of the liver, mature cells remain capable of division to allow growth or regeneration after injury.

In contrast to these patterns, other types of cells either cannot divide or are prevented from dividing by certain molecules produced by nearby cells. As a result, in the adult organism, some tissues have a greatly reduced capacity to renew damaged or diseased cells. Examples of such tissues include heart muscle, nerve cells of the central nervous system, and lens cells in mammals. Maintenance and repair of these cells is limited to replacing intracellular components rather than replacing entire cells.

Duplication of the Genetic Material

Before a cell can divide, it must accurately and completely duplicate the genetic information encoded in its DNA in order for its progeny cells to function and survive. This is a complex problem because of the great length of DNA molecules. Each human chromosome consists of a long double spiral, or helix, each strand of which consists of more than 100 million nucleotides.

The duplication of DNA is called DNA replication, and it is initiated by complex enzymes called DNA polymerases. These progress along the molecule, reading the sequences of nucleotides that are linked together to make DNA chains. Each strand of the DNA double helix, therefore, acts as a template specifying the nucleotide structure of a new growing chain. After replication, each of the two daughter DNA double helices consists of one parental DNA strand wound around one newly synthesized DNA strand.

In order for DNA to replicate, the two strands must be unwound from each other. Enzymes called helicases unwind the two DNA strands, and additional proteins bind to the separated strands to stabilize them and prevent them from pairing again. In addition, a remarkable class of enzyme called DNA topoisomerase removes the helical twists by cutting either one or both strands and then resealing the cut. These enzymes can also untangle and unknot DNA when it is tightly coiled into a chromatin fibre.

In the circular DNA of prokaryotes, replication starts at a unique site called the origin of replication and then proceeds in both directions around the molecule until the two processes meet, producing two daughter molecules. In rapidly growing prokaryotes, a second round of replication can start before the first has finished. The situation in eukaryotes is more complicated, as replication moves more slowly than in prokaryotes. At 500 to 5,000 nucleotides per minute (versus 100,000 nucleotides per minute in prokaryotes), it would take a human chromosome about a month to replicate if started at a single site. Actually, replication begins at many sites on the long chromosomes of animals, plants, and fungi. Distances between adjacent initiation sites are not always the same; for example, they are closer in the rapidly dividing embryonic cells of frogs or flies than in adult cells of the same species.

Accurate DNA replication is crucial to ensure that daughter cells have exact copies of the genetic information for synthesizing proteins. Accuracy is achieved by a "proofreading" ability of the DNA polymerase itself. It can erase its own errors and then synthesize anew. There are also repair systems that correct genetic damage to DNA. For example, the incorporation of an incorrect nucleotide, or damage caused by mutagenic agents, can be corrected by cutting out a section of the daughter strand and recopying the parental strand.

Cell Division

Mitosis and Cytokinesis

In eukaryotes the processes of DNA replication and cell division occur at different times of the cell division cycle. During cell division, DNA condenses to form short, tightly coiled, rodlike chromosomes. Each chromosome then splits longitudinally, forming two identical chromatids. Each pair of chromatids is divided between the two daughter cells during mitosis, or division of the nucleus, a process in which the chromosomes are propelled by attachment to a bundle of microtubules called the mitotic spindle.

Mitosis can be divided into five phases. In prophase the mitotic spindle forms and the chromosomes condense. In prometaphase the nuclear envelope breaks down (in many but not all eukaryotes) and the chromosomes attach to the mitotic spindle. Both chromatids of each chromosome attach to the spindle at a specialized chromosomal region called the kinetochore. In metaphase the condensed chromosomes align in a plane across the equator of the mitotic spindle. Anaphase follows as the separated chromatids move abruptly toward opposite spindle poles. Finally, in telophase a new nuclear envelope forms around each set of unraveling chromatids.

An essential feature of mitosis is the attachment of the chromatids to opposite poles of the mitotic spindle. This ensures that each of the daughter cells will receive a complete set of chromosomes. The mitotic spindle is composed of microtubules, each of which is a tubular assembly of molecules of the protein tubulin. Some microtubules extend from one spindle pole to the other, while a second class extends from one spindle pole to a chromatid. Microtubules can grow or shrink by the addition or removal of tubulin molecules. The shortening of spindle microtubules at anaphase propels attached chromatids to the spindle poles, where they unravel to form new nuclei.

The two poles of the mitotic spindle are occupied by centrosomes, which organize the microtubule arrays. In animal cells each centrosome contains a pair of cylindrical centrioles, which are themselves composed of complex arrays of microtubules. Centrioles duplicate at a precise time in the cell division cycle, usually close to the start of DNA replication.

After mitosis comes cytokinesis, the division of the cytoplasm. This is another process in which animal and plant cells differ. In animal cells cytokinesis is achieved through the constriction of the cell by a ring of contractile microfilaments consisting of actin and myosin, the proteins involved in muscle contraction and other forms of cell movement. In plant cells the cytoplasm is divided by the formation of a new cell wall, called the cell plate, between the two daughter cells. The cell plate arises from small Golgi-derived vesicles that coalesce in a plane across the equator of the late telophase spindle to form a disk-shaped structure. In this process, each vesicle contributes its membrane to the forming cell membranes and its matrix contents to the forming cell wall. A second set of vesicles extends the edge of the cell plate until it reaches and fuses with the sides of the parent cell, thereby completely separating the two new daughter cells. At this point, cellulose synthesis commences, and the cell plate becomes a primary cell wall.

Meiosis

A specialized division of chromosomes called meiosis occurs during the formation of the reproductive cells, or gametes, of sexually reproducing organisms. Gametes such as ova, sperm, and pollen begin as germ cells, which, like other types of cells, have two copies of each gene in their nuclei. The chromosomes composed of these matching genes are called homologs. During DNA replication, each chromosome duplicates into two attached chromatids. The homologous chromosomes are then separated to opposite poles of the meiotic spindle by microtubules similar to those of the mitotic spindle. At this stage in the meiosis of germ cells, there is a crucial difference from the mitosis of other cells. In meiosis the two chromatids making up each chromosome remain together, so that whole chromosomes are separated from their homologous partners. Cell division then occurs, followed by a second division that resembles mitosis more closely in that it separates the two chromatids of each remaining chromosome. In this way, when meiosis is complete, each mature gamete receives only one copy of each gene instead of the two copies present in other cells.

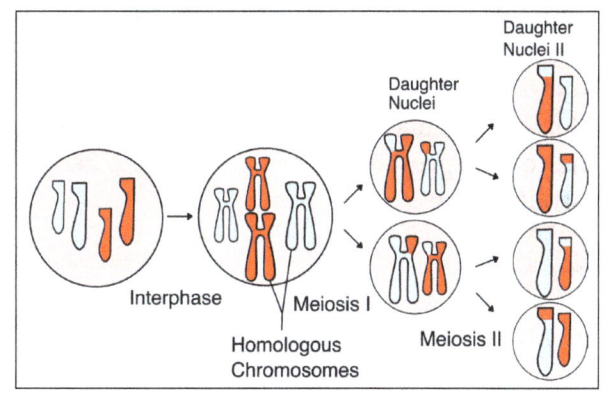

The formation of gametes (sex cells) occurs during the process of meiosis.

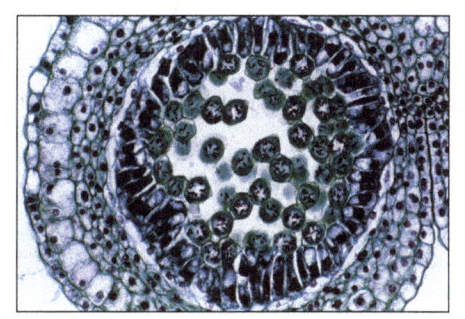

Meiosis; pollen Cross section of a plant anther showing meiosis occurring in pollen grain cells.

The Cell Division Cycle

In prokaryotes, DNA synthesis can take place uninterrupted between cell divisions, and new cycles of DNA synthesis can begin before previous cycles have finished. In contrast, eukaryotes duplicate their DNA exactly once during a discrete period between cell divisions. This period is called the S (for synthetic) phase. It is preceded by a period called G_1 (meaning "first gap") and followed by a period called G_2, during which nuclear DNA synthesis does not occur.

The four periods G_1, S, G_2, and M (for mitosis) make up the cell division cycle. The cell cyclecharacteristically lasts between 10 and 20 hours in rapidly proliferating adult cells, but it can be arrested for weeks or months in quiescent cells or for a lifetime in neurons of the brain. Prolonged arrest of this type usually occurs during the G_1 phase and is sometimes referred to as G_0. In contrast, some embryonic cells, such as those of fruit flies (vinegar flies), can complete entire cycles and divide in only 11 minutes. In these exceptional cases, G_1 and G_2 are undetectable, and mitosis alternates with DNA synthesis. In addition, the duration of the S phase varies dramatically. The fruit fly embryo takes only four minutes to replicate its DNA, compared with several hours in adult cells of the same species.

Controlled Proliferation

Several studies have identified the transition from the G1 to the S phase as a crucial control point of the cell cycle. Stimuli are known to cause resting cells to proliferate by inducing them to leave G1 and begin DNA synthesis. These stimuli, called growth factors, are naturally occurring proteins specific to certain groups of cells in the body. They include nerve growth factor, epidermal growth factor, and platelet-derived growth factor. Such factors may have important roles in the healing of wounds as well as in the maintenance and growth of normal tissues. Many growth factors are known to act on the external membrane of the cell, by interacting with specialized protein receptor molecules. These respond by triggering further cellular changes, including an increase in calcium levels that makes the cell interior more alkaline and the addition of phosphate groups to the amino acid tyrosine in proteins. The complex response of cells to growth factors is of fundamental importance to the control of cell proliferation.

Failure of Proliferation Control

Cancer can arise when the controlling factors over cell growth fail and allow a cell and its descendants to keep dividing at the expense of the organism. Studies of viruses that transform cultured cells and thus lead to the loss of control of cell growth have provided insight into the mechanisms that drive the formation of tumours. Transformed cells may differ from their normal progenitors by continuing to proliferate at very high densities, in the absence of growth factors, or in the absence of a solid substrate for support.

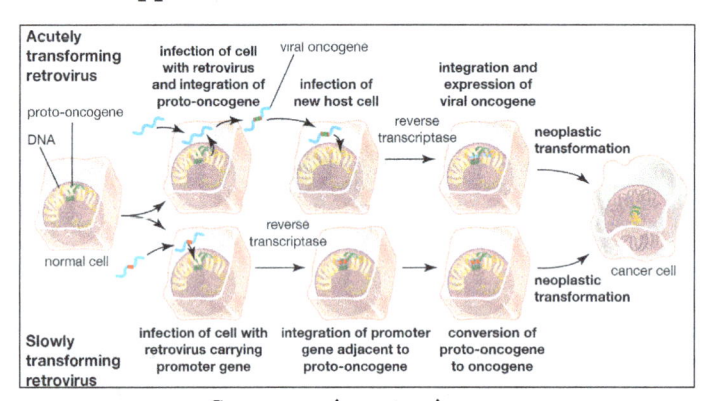

Cancer-causing retroviruses.

Retroviral insertion can convert a proto-oncogene, integral to the control of cell division, into an oncogene, the agent responsible for transforming a healthy cell into a cancer cell. An acutely transforming retrovirus (shown at top), which produces tumours within weeks of infection, incorporates genetic material from a host cell into its own genome upon infection, forming a viral oncogene. When the viral oncogene infects another cell, an enzyme called reverse transcriptase copies the single-stranded genetic material into double-stranded DNA, which is then integrated into the cellular genome. A slowly transforming retrovirus (shown at bottom), which requires months to elicit tumour growth, does not disrupt cellular function through the insertion of a viral oncogene. Rather, it carries a promoter gene that is integrated into the cellular genome of the host cell next to or within a proto-oncogene, allowing conversion of the proto-oncogene to an oncogene.

Major advances in the understanding of growth control have come from studies of the viral genes that cause transformation. These viral oncogenes have led to the identification of related cellular genes called protooncogenes. Protooncogenes can be altered by mutation or epigenetic modification, which converts them into oncogenes and leads to cell transformation. Specific oncogenes are activated in particular human cancers. For example, an oncogene called RAS is associated with many epithelial cancers, while another, called MYC, is associated with leukemias.

An interesting feature of oncogenes is that they may act at different levels corresponding to the multiple steps seen in the development of cancer. Some oncogenes immortalize cells so that they divide indefinitely, whereas normal cells die after a limited number of generations. Other oncogenes transform cells so that they grow in the absence of growth factors. A combination of these two functions leads to loss of proliferation control, whereas each of these functions on its own cannot. The mode of action of oncogenes also provides important clues to the nature of growth control and cancer. For example, some oncogenes are known to encode receptors for growth factors that may cause continuous proliferation in the absence of appropriate growth factors.

Loss of growth control has the added consequence that cells no longer repair their DNA effectively, and thus aberrant mitoses occur. As a result, additional mutations arise that subvert a cell's normal constraints to remain in its tissue of origin. Epithelial tumour cells, for example, acquire the ability to cross the basal lamina and enter the bloodstream or lymphatic system, where they migrate to other parts of the body, a process called metastasis. When cells metastasize to distant tissues, the tumour is described as malignant, whereas prior to metastasis a tumour is described as benign.

Cell Differentiation

Adult organisms are composed of a number of distinct cell types. Cells are organized into tissues, each of which typically contains a small number of cell types and is devoted to a specific physiological function. For example, the epithelial tissue lining the small intestine contains columnar absorptive cells, mucus-secreting goblet cells, hormone-secreting endocrine cells, and enzyme-secreting Paneth cells. In addition, there exist undifferentiated dividing cells that lie in the crypts between the intestinal villi and serve to replace the other cell types when they become damaged or worn out. Another example of a differentiated tissue is the skeletal tissue of a long bone, which contains osteoblasts (large cells that synthesize bone) in the outer sheath and osteocytes (mature bone cells) and osteoclasts (multinucleate cells involved in bone remodeling) within the matrix.

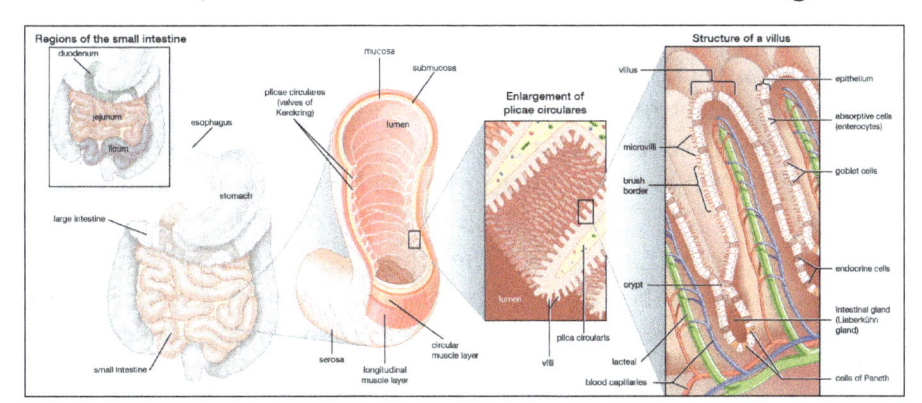

The small intestine contains many distinct types of cells, each of which serves a specific function.

In general, the simpler the overall organization of the animal, the fewer the number of distinct cell types that they possess. Mammals contain more than 200 different cell types, whereas simple invertebrate animals may have only a few different types. Plants are also made up of differentiated cells, but they are quite different from the cells of animals. For example, a leaf in a higher plant is covered with a cuticle layer of epidermal cells. Among these are pores composed of two specialized cells, which regulate gaseous exchange across the epidermis. Within the leaf is the mesophyll, a spongy tissue responsible for photosynthetic activity. There are also veins composed of xylem elements, which transport water up from the soil, and phloem elements, which transport products of photosynthesis to the storage organs.

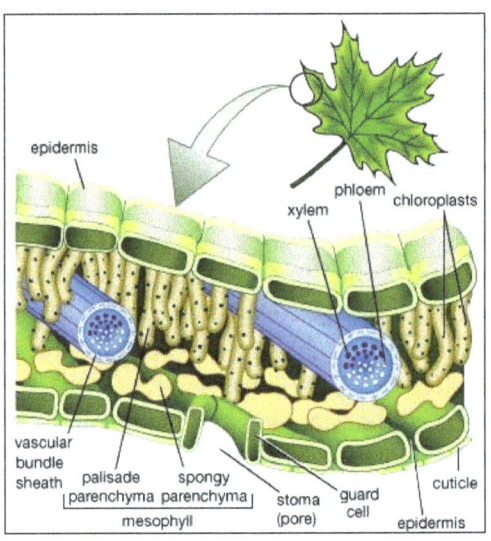

Structures of a leaf.

The epidermis is often covered with a waxy protective cuticle that helps prevent water loss from inside the leaf. Oxygen, carbon dioxide, and water enter and exit the leaf through pores (stomata) scattered mostly along the lower epidermis. The stomata are opened and closed by the contraction and expansion of surrounding guard cells. The vascular, or conducting, tissues are known as xylem and phloem; water and minerals travel up to the leaves from the roots through the xylem, and sugars made by photosynthesis are transported to other parts of the plant through the phloem. Photosynthesis occurs within the chloroplast-containing mesophyll layer.

The various cell types have traditionally been recognized and classified according to their appearance in the light microscope following the process of fixing, processing, sectioning, and staining tissues that is known as histology. Classical histology has been augmented by a variety of more discriminating techniques. Electron microscopy allows for higher magnifications. Histochemistry involves the use of coloured precipitating substrates to stain particular enzymes in situ. Immunohistochemistry uses specific antibodies to identify particular substances, usually proteins or carbohydrates, within cells. In situ hybridization involves the use of nucleic acid probes to visualize the location of specific messenger RNAs (mRNA). These modern methods have allowed the identification of more cell types than could be visualized by classical histology, particularly in the brain, the immune system, and among the hormone-secreting cells of the endocrine system.

The Differentiated State

The biochemical basis of cell differentiation is the synthesis by the cell of a particular set of proteins, carbohydrates, and lipids. This synthesis is catalyzed by proteins called enzymes. Each enzyme in turn is synthesized in accordance with a particular gene, or sequence of nucleotides in the DNA of the cell nucleus. A particular state of differentiation, then, corresponds to the set of genes that is expressed and the level to which it is expressed.

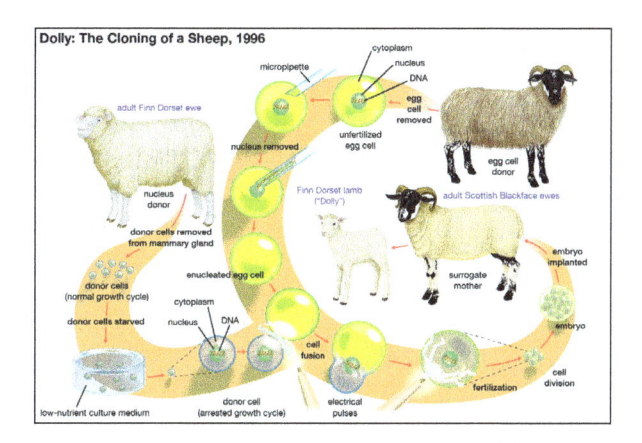

Dolly the sheep was successfully cloned in 1996 by fusing the nucleus from a mammary-gland cell of a Finn Dorset ewe into an enucleated egg cell taken from a Scottish Blackface ewe. Carried to term in the womb of another Scottish Blackface ewe, Dolly was a genetic copy of the Finn Dorset ewe.

It is believed that all of an organism's genes are present in each cell nucleus, no matter what the cell type, and that differences between tissues are not due to the presence or absence of certain genes but are due to the expression of some and the repression of others. In animals the best evidence for retention of the entire set of genes comes from whole animal cloning experiments in which the nucleus of a differentiated cell is substituted for the nucleus of a fertilized egg. In many species this can result in the development of a normal embryo that contains the full range of body parts and cell types. Likewise, in plants it is often possible to grow complete embryos from individual cells in tissue culture. Such experiments show that any nucleus has the genetic information required for the growth of a developing organism, and they strongly suggest that, for most tissues, cell differentiation arises from the regulation of genetic activity rather than the removal or destruction of unwanted genes. The only known exception to this rule comes from the immune system, where segments of DNA in developing white blood cells are slightly rearranged, producing a wide variety of antibody and receptor molecules.

At the molecular level there are many ways in which the expression of a gene can be differentially regulated in different cell types. There may be differences in the copying, or transcription, of the gene into RNA; in the processing of the initial RNA transcript into mRNA; in the control of mRNA movement to the cytoplasm; in the translation of mRNA to protein; or in the stability of mRNA. However, the control of transcription has the most influence over gene expression and has received the most detailed analysis.

The DNA in the cell nucleus exists in the form of chromatin, which is made up of DNA bound to histones (simple alkaline proteins) and other nonhistone proteins. Most of the DNA is complexed into repeating structures called nucleosomes, each of which contains eight molecules of histone. Active genes are found in parts of the DNA where the chromatin has an "open" configuration, in which regulatory proteins are able to gain access to the DNA. The degree to which the chromatin opens depends on chemical modifications of the outer parts of the histone molecules and on the presence or absence of particular nonhistone proteins. Transcriptional control is exerted with the help of regulatory sequences that are found associated with a gene, such as the promoter sequence, a region near the start of the gene, and enhancer sequences, regions that lie elsewhere within the DNA that augment the activity of enzymes involved in the process of transcription. Whether or not transcription occurs depends on the binding of transcription factors to these regulatory sequences. Transcription factors are proteins that usually possess a DNA-binding region, which recognizes the specific regulatory sequence in the DNA, and an effector region, which activates or inhibits transcription. Transcription factors often work by recruiting enzymes that add modifications (e.g., acetyl groups or methyl groups) to or remove modifications from the outer parts of the histone molecules. This controls the folding of the chromatin and the accessibility of the DNA to RNA polymerase and other transcription factors.

In general, it requires several transcription factors working in combination to activate a gene. For example, the chicken delta 1 crystallin gene, normally expressed only in the lens of the eye, has a promoter that contains binding sites for two activating transcription factors and an enhancer that contains binding sites for two other activating transcription factors. There is also an additional enhancer site that can bind either an activator (deltaEF3) or a repressor (deltaEF1). Successful transcription requires that all these sites are occupied by the correct transcription factors.

Fully differentiated cells are qualitatively different from one another. States of terminal differentiation are stable and persistent, both in the lifetime of the cell and in successive cell generations (in the case of differentiated types that are capable of continued cell division). The inherent stability of the differentiated state is maintained by various processes, including feedback activation of genes by their own products and repression of inactive genes. Chromatin structure may be important in maintaining states of differentiation, although it is still unclear whether this can be maintained during DNA replication, which involves temporary removal of chromosomal proteins and unwinding of the DNA double helix.

A type of differentiation control that is maintained during DNA replication is the methylation of DNA, which tends to recruit histone deacetylases and hence close up the structure of the chromatin. DNA methylation occurs when a methyl group is attached to the exterior, or sugar-phosphate side, of a cytosine (C) residue. Cytosine methylation occurs only on a C nucleotide when it is connected to a G (guanine) nucleotide on the same strand of DNA. These nucleotide pairings are called CG dinucleotides. One class of DNA methylase enzyme can introduce new methylations when required, whereas another class, called maintenance methylases, methylates CG dinucleotides in the DNA double helix only when the CG of the complementary strand is already methylated. Each time the methylated DNA is replicated, the old strand has the methyl groups and the new strand does not. The maintenance methylase will then add methyl groups to all the CGs opposite the existing methyl groups to restore a fully methylated double helix. This mechanism guarantees stability of the DNA methylation pattern, and hence the differentiated state, during the processes of DNA replication and cell division.

The Process of Differentiation

Differentiation from visibly undifferentiated precursor cells occurs during embryonic development, during metamorphosis of larval forms, and following the separation of parts in asexual reproduction. It also takes place in adult organisms during the renewal of tissues and the regeneration of missing parts. Thus, cell differentiation is an essential and ongoing process at all stages of life.

The visible differentiation of cells is only the last of a progressive sequence of states. In each state, the cell becomes increasingly committed toward one type of cell into which it can develop. States of commitment are sometimes described as "specification" to represent a reversible type of commitment and as "determination" to represent an irreversible commitment. Although states of specification and determination both represent differential gene activity, the properties of embryonic cells are not necessarily the same as those of fully differentiated cells. In particular, cells in specification states are usually not stable over prolonged periods of time.

Two mechanisms bring about altered commitments in the different regions of the early embryo: cytoplasmic localization and induction. Cytoplasmic localization is evident in the earliest stages of development of the embryo. During this time, the embryo divides without growth, undergoing cleavage divisions that produce separate cells called blastomeres. Each blastomere inherits a certain region of the original egg cytoplasm, which may contain one or more regulatory substances called cytoplasmic determinants. When the embryo has become a solid mass of blastomeres (called a morula), it generally consists of two or more differently committed cell populations—a result of the blastomeres having incorporated different cytoplasmic determinants. Cytoplasmic determinants may consist of mRNA or protein in a particular state of activation. An example of the influence of a cytoplasmic determinant is a receptor called Toll, located in the membranes of Drosophila (fruit fly) eggs. Activation of Toll ensures that the blastomeres will develop into ventral (underside) structures, while blastomeres containing inactive Toll will produce cells that will develop into dorsal (back) structures.

In induction, the second mechanism of commitment, a substance secreted by one group of cells alters the development of another group. In early development, induction is usually instructive; that is, the tissue assumes a different state of commitment in the presence of the signal than it would in the absence of the signal. Inductive signals often take the form of concentration gradients of substances that evoke a number of different responses at different concentrations. This leads to the formation of a sequence of groups of cells, each in a different state of specification. For example, in Xenopus (clawed frog) the early embryo contains a signaling centre called the organizer that secretes inhibitors of bone morphogenetic proteins (BMPs), leading to a ventral-to-dorsal (belly-to-back) gradient of BMP activity. The activity of BMP in the ventral region of the embryo suppresses the expression of transcription factors involved in the formation of the central nervous system and segmented muscles. Suppression ensures that these structures are formed only on the dorsal side, where there is decreased activity of BMP.

The final stage of differentiation often involves the formation of several types of differentiated cells from one precursor or stem cell population. Terminal differentiation occurs not only in embryonic development but also in many tissues in postnatal life. Control of this process depends on a system of lateral inhibition in which cells that are differentiating along a particular pathway send out signals that repress similar differentiation by their neighbours. For example, in the developing central nervous system of vertebrates, neurons arise from a simple tube of neuroepithelium, the cells of which possess a surface receptor called Notch. These cells also possess another cell surface molecule called Delta that can bind to and activate Notch on adjacent cells. Activation of Notch initiates a cascade of intracellular events that results in suppression of Delta production and suppression of neuronal differentiation. This means that the neuroepithelium generates only a few cells with high expression of Delta surrounded by a larger number of cells with low expression of Delta. High Delta production and low Notch activation makes the cells develop into neurons. Low

Delta production and high Notch activation makes the cells remain as precursor cells or become glial (supporting) cells. A similar mechanism is known to produce the endocrine cells of the pancreas and the goblet cells of the intestinal epithelium. Such lateral inhibition systems work because cells in a population are never quite identical to begin with. There are always small differences, such as in the number of Delta molecules displayed on the cell surface. The mechanism of lateral inhibition amplifies these small differences, using them to bring about differential gene expression that leads to stable and persistent states of cell differentiation.

Errors in Differentiation

Three classes of abnormal cell differentiation are dysplasia, metaplasia, and anaplasia. Dysplasia indicates an abnormal arrangement of cells, usually arising from a disturbance in their normal growth behaviour. Some dysplasias are precursor lesions to cancer, whereas others are harmless and regress spontaneously. For example, dysplasia of the uterine cervix, called cervical intraepithelial neoplasia (CIN), may progress to cervical cancer. It can be detected by cervical smear cytology tests (Pap smears).

Metaplasia is the conversion of one cell type into another. In fact, it is not usually the differentiated cells themselves that change but rather the stem cell population from which they are derived. Metaplasia commonly occurs where chronic tissue damage is followed by extensive regeneration. For example, squamous metaplasia of the bronchi occurs when the ciliated respiratory epithelial cells of people who smoke develop into squamous, or flattened, cells. In intestinal metaplasia of the stomach, patches resembling intestinal tissue arise in the gastric mucosa, often in association with gastric ulcers. Both of these types of metaplasia may progress to cancer.

Anaplasia is a loss of visible differentiation that can occur in advanced cancer. In general, early cancers resemble their tissue of origin and are described and classified by their pattern of differentiation. However, as they develop, they produce variants of more abnormal appearance and increased malignancy. Finally, a highly anaplastic growth can occur, in which the cancerous cells bear no visible relation to the parent tissue.

Cell Theory

Cell theory is a proposed and widely accepted view of how most life on Earth functions. According to the theory, all organisms are made of cells. Groups of cells create tissues, organs, and organisms. Further, cells can only arise from other cells. These are the main tenants of cell theory.

Before the invention of advanced microscopes, microorganisms were unknown, and it was assumed that individuals were the basic units of life. this view began to change thanks to the microscope. Microscopes allowed early scientists to view and postulate about the cells they could see. Even with a microscope, it is not always possible to see the exact functioning of a cell. Scientists formulated a general theory of how cells work that is very simple.

This theory revolves around the fact that no matter what type of organism we view under a microscope, the organisms are clearly divided into a number of different cells. Some cells are very large, such as a frog egg. Other cells, such as some bacterial cells, are so small we can barely see them with a normal light microscope. Viruses, which may or may not be living, are the only forms of reproducing DNA or RNA which are not always contained within a cell.

3 Parts of Cell Theory

Cell theory has three major hypotheses:

- First, all organisms are made of cells.
- Second, cells are the fundamental building blocks used to create tissues, organs, and entire functioning organisms.
- The third, and probably most important part of the theory is that cells can only arise from other cells.

Thus, all organisms start as single cells. These cells grow, divide through mitosis, and develop into multi-celled organisms. Mitosis is a form of cell division that produces identical cells. These cells can then differentiate when given different signals to produce different types of tissues and organs. This is how large and complex organisms are made. Single-celled organisms divide as well, but when they divide, the cells separate into two new individuals. This is known as asexual reproduction.

Cell Theory Examples

Single-Celled Organisms

Single-celled organisms are a great way to study cell theory. With modern microscopes, the processes behind cell theory can easily be viewed and studied. A great example of watching cell theory in action can be accomplished by putting a drop of pond water under a microscope. Below is a picture of two Euglena organisms, seen just after reproduction.

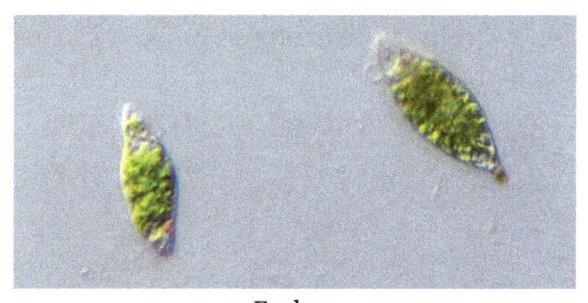

Euglena

Minutes before, these two cells were one. Euglena reproduces through simple cellular division. The DNA in the parent organism is duplicated, as are the internal organelles. Then, the large cell divides into two equally-sized smaller cells, as seen in the picture. These two cells are now independent organisms. Each will try to survive, grow, and eventually reproduce again.

In Plants

Cells were first discovered in plants. Plants, have large structures called *cell walls*, which enable the plant to remain rigid. These cell walls are easily visible, even with the first microscope invented in 1665. Robert Hooke, the man who first identified cells, did so using a simple microscope aimed at a thin slice of cork. He drew what he saw, and published it in a book about microscopy.

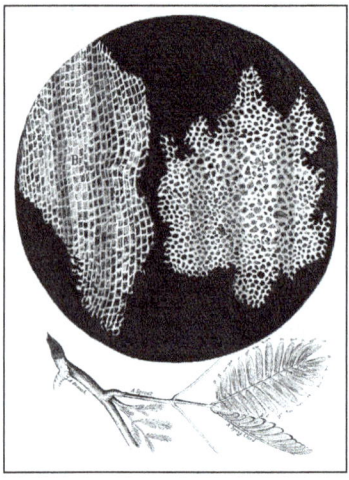

A thin slice of cork.

As you can see, Hooke was clearly looking at cells. In fact, with a better microscope, he likely could have seen the cells in action and the many organelles inside. Instead, Hooke did not come to the immediate conclusion that all organisms were made of cells. He assumed the structures were limited to the tissues of plants. It was not until the 1840s that cell theory would be largely accepted by science.

In Animals

Scientist Theodor Schwann presented evidence that animals, like plants, were also fundamentally composed of different types of cells. Modern microscopy techniques allow scientists a much more comprehensive and accurate view of cells compared to early scientists. Below is a scanning electron micrograph of red blood cells. It distinctly shows how our red blood cells are separate, functional units of the human body.

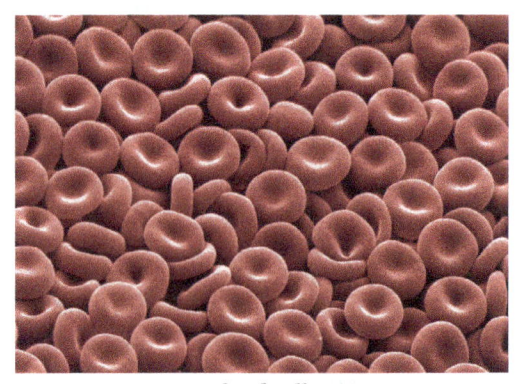

Blood cell.

Like red blood cells, every part of the body is composed of different types of cells. According to cell theory, all of these cells are derived from the zygote, which is a single cell that results from the fertilization of an egg with a sperm. This cell then divides, replicates, and begins to differentiate into the many different cell types of the body. Eventually, a fully-functional organism is formed.

Other Organisms

Cells are the basic building blocks of all life on Earth. This is true of fungi, the only kingdomnot yet covered. In fact, fungi are a sort of intermediate between plants and animals. While they lack the sun-harvesting chloroplasts of plants, they do have cell walls. However, there is one form of life which does not strictly adhere to cell theory.

Viruses are small DNA or RNA particles, surrounded by a protective protein coating. Many scientists do not consider viruses a living organism, and thus it is okay that they do not conform to the typical cell theory. Other scientists consider them living, but suggest they are an exception to cell theory. For viruses to reproduce they must infect a host cell. Only by using the host cell's machinery can a virus replicate its genetic code and the proteins needed to create new virus particles.

Cell Types

Because of the millions of diverse species of life on Earth, which grow and change gradually over time, there are countless differences between the countless extant types of cells.

Eukaryotic Cell

Eukaryotic cells are cells that contain a membrane-bound nucleus, a structural feature that is not present in bacterial or archaeal cells. In addition to the nucleus, eukaryotic cells are characterized by numerous membrane-bound organelles such as the endoplasmic reticulum, Golgi apparatus, chloroplasts, mitochondria, and others.

We began to consider the Design Challenge of making cells larger than a small bacterium—more precisely, growing cells to sizes at which, in the eyes of natural selection, relying on diffusion of substances for transport through a highly viscous cytosol comes with inherent functional trade-offs that offset most selective benefits of getting larger. Bacterial cell structure, we discovered some morphological features of large bacteria that allow them to effectively overcome diffusion-limited size barriers (e.g., filling the cytoplasm with a large storage vacuole maintains a small volume for metabolic activity that remains compatible with diffusion-driven transport).

As we transition our focus to eukaryotic cells, we want you to approach the study by constantly returning to the Design Challenge. We will cover a large number of subcellular structures that are unique to eukaryotes, and you will certainly be expected to know the names of these structures or organelles, to associate them with one or more "functions", and to identify them on a canonical cartoon representation of a eukaryotic cell. This memorization exercise is necessary but not sufficient. We will also ask you to start thinking a bit deeper about some of the functional and evolutionary costs and benefits (trade-offs) of both evolving eukaryotic cells and various eukaryotic organelles, as well as how a eukaryotic cell might coordinate the functions of different organelles.

Our hypotheses may sometimes come in the form of statements like, "Thing A exists because of rationale B." To be completely honest, however, in many cases, we don't actually know all of the selective pressures that led to the creation or maintenance of certain cellular structures, and the likelihood that one explanation will fit all cases is slim in biology. The causal linkage/relationship implied by the use of terms like "because" should be treated as good hypotheses rather than objective, concrete, undisputed, factual knowledge. We want you to understand these hypotheses and to be able to discuss the ideas presented in class, but we also want you to indulge your own curiosity and to begin thinking critically about these ideas yourself. Try using the Design Challenge rubric to explore some of your ideas. In the following, we will try to seed questions to encourage this activity.

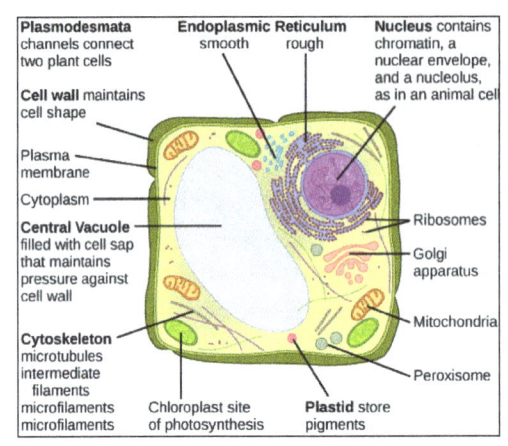

Figure: These figures show the major organelles and other cell components of (a) a typical animal cell and (b) a typical eukaryotic plant cell. The plant cell has a cell wall, chloroplasts, plastids, and a central vacuole—structures not found in animal cells. Plant cells do not have lysosomes or centrosomes.

The Plasma Membrane

Like bacteria and archaea, eukaryotic cells have a plasma membrane, a phospholipid bilayer with embedded proteins that separates the internal contents of the cell from its surrounding environment. The plasma membrane controls the passage of organic molecules, ions, water, and oxygen into and out of the cell. Wastes (such as carbon dioxide and ammonia) also leave the cell by passing through the plasma membrane, usually with some help of protein transporters.

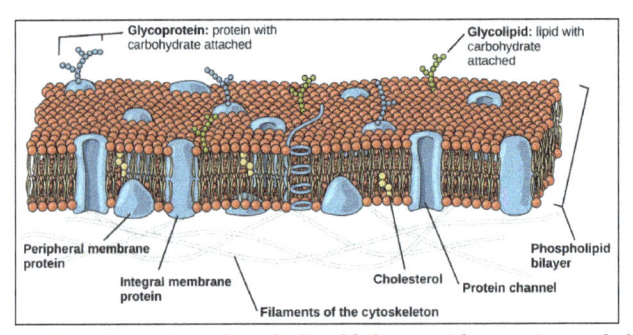

Figure: The eukaryotic plasma membrane is a phospholipid bilayer with proteins and cholesterol embedded in it.

As discussed in the context of bacterial cell membranes, the plasma membranes of eukaryotic cells may also adopt unique conformations. For instance, the plasma membrane of cells that, in multicellular organisms, specialize in absorption are often folded into fingerlike projections called microvilli (singular = microvillus);. The "folding" of the membrane into microvilli effectively increases the surface area for absorption while minimally impacting the cytosolic volume. Such cells can be found lining the small intestine, the organ that absorbs nutrients from digested food.

People with celiac disease have an immune response to gluten, a protein found in wheat, barley, and rye. The immune response damages microvilli. As a consequence, afflicted individuals have an impaired ability to absorb nutrients. This can lead to malnutrition, cramping, and diarrhea.

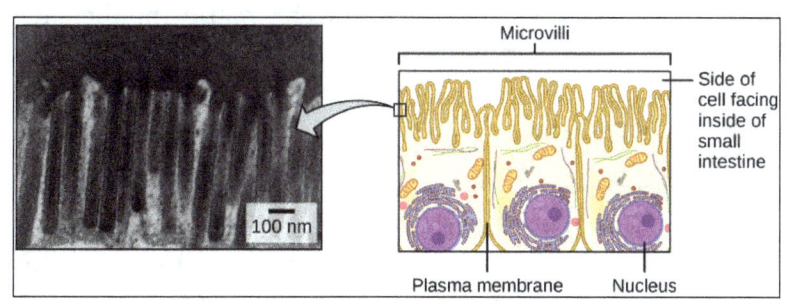

Figure: Microvilli, shown here as they appear on cells lining the small intestine, increase the surface area available for absorption. These microvilli are only found on the area of the plasma membrane that faces the cavity from which substances will be absorbed.

The Cytoplasm

The cytoplasm refers to the entire region of a cell between the plasma membrane and the nuclear envelope. It is composed of organelles suspended in the gel-like cytosol, the cytoskeleton, and various chemicals. Even though the cytoplasm consists of 70 to 80 percent water, it nevertheless has a semisolid consistency. It is crowded in there. Proteins, simple sugars, polysaccharides, amino acids, nucleic acids, fatty acids, ions and many other water-soluble molecules are all competing for space and water.

The Nucleus

Typically, the nucleus is the most prominent organelle in a cell when viewed through a microscope. The nucleus (plural = nuclei) houses the cell's DNA. Let's look at it in more detail.

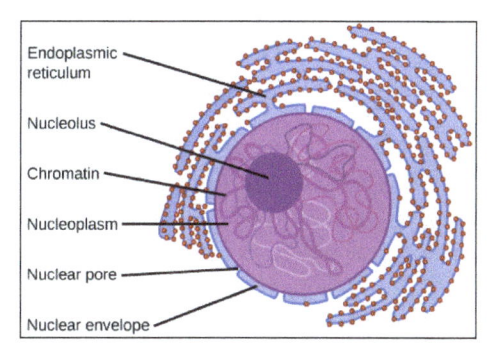

The nucleus stores chromatin (DNA plus proteins) in a gel-like substance called the nucleoplasm. The nucleolus is a condensed region of chromatin where ribosome synthesis occurs. The boundary of the nucleus is called the nuclear envelope. It consists of two phospholipid bilayers: An outer membrane and an inner membrane. The nuclear membrane is continuous with the endoplasmic reticulum. Nuclear pores allow substances to enter and exit the nucleus.

The Nuclear Envelope

The nuclear envelope, a structure that constitutes the outermost boundary of the nucleus, is a double-membrane—both the inner and outer membranes of the nuclear envelope are phospholipid bilayers. The nuclear envelope is also punctuated with protein-based pores that control the passage of ions, molecules, and RNA between the nucleoplasm and cytoplasm. The nucleoplasm is the semisolid fluid inside the nucleus where we find the chromatin and the nucleolus, a condensed region of chromatin where ribosome synthesis occurs.

Chromatin and Chromosomes

To understand chromatin, it is helpful to first consider chromosomes. Chromosomes are structures within the nucleus that are made up of DNA, the hereditary material. You may remember that in bacteria and archaea, DNA is typically organized into one or more circular chromosome(s). In eukaryotes, chromosomes are linear structures. Every eukaryotic species has a specific number of chromosomes in the nuclei of its cells. In humans, for example, the chromosome number is 23, while in fruit flies, it is 4.

Chromosomes are only clearly visible and distinguishable from one another by visible optical microscopy when the cell is preparing to divide and the DNA is tightly packed by proteins into easily distinguishable shapes. When the cell is in the growth and maintenance phases of its life cycle, numerous proteins are still associated with the nucleic acids, but the DNA strands more closely resemble an unwound, jumbled bunch of threads. The term chromatin is used to describe chromosomes (the protein-DNA complexes) when they are both condensed and decondensed.

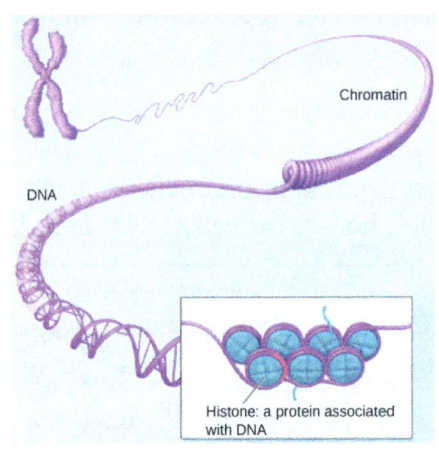

Figure: (a) This image shows various levels of the organization of chromatin (DNA and protein). (b) This image shows paired chromosomes. Credit (b): modification of work by NIH; scale-bar data from Matt Russell.

The Nucleolus

Some chromosomes have sections of DNA that encode ribosomal RNA. A darkly staining area within the nucleus called the nucleolus (plural = nucleoli) aggregates the ribosomal RNA with associated proteins to assemble the ribosomal subunits that are then transported out to the cytoplasm through the pores in the nuclear envelope.

Ribosomes

Ribosomes are the cellular structures responsible for protein synthesis. When viewed through an electron microscope, ribosomes appear either as clusters (polyribosomes) or single, tiny dots that

Cell: The Fundamental Unit of Life

float freely in the cytoplasm. They may be attached to the cytoplasmic side of the plasma membrane or the cytoplasmic side of the endoplasmic reticulum and the outer membrane of the nuclear envelope.

Electron microscopy has shown us that ribosomes, which are large complexes of protein and RNA, consist of two subunits, aptly called large and small (figure below). Ribosomes receive their "instructions" for protein synthesis from the nucleus, where the DNA is transcribed into messenger RNA (mRNA). The mRNA travels to the ribosomes, which translate the code provided by the sequence of the nitrogenous bases in the mRNA into a specific order of amino acids in a protein. This is covered in greater detail in the section covering the process of translation.

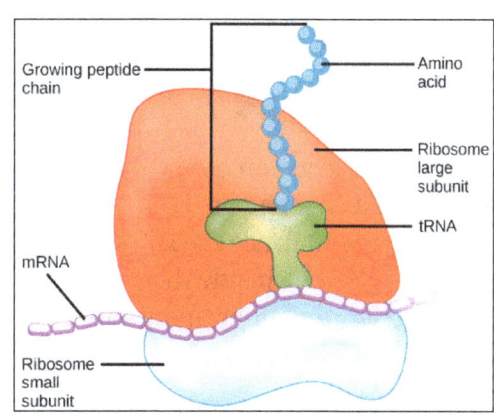

Figure: Ribosomes are made up of a large subunit (top) and a small subunit (bottom). During protein synthesis, ribosomes assemble amino acids into proteins.

Mitochondria

Mitochondria (singular = mitochondrion) are often called the "powerhouses" or "energy factories" of a cell because they are the primary site of metabolic respiration in eukaryotes. Depending on the species and the type of mitochondria found in those cells, the respiratory pathways may be anaerobic or aerobic. By definition, when respiration is aerobic, the terminal electron is oxygen; when respiration is anaerobic, a compound other than oxygen functions as the terminal electron acceptor. In either case, the result of these respiratory processes is the production of ATP via oxidative phosphorylation, hence the use of terms "powerhouse" and/or "energy factory" to describe this organelle. Nearly all mitochondria also possess a small genome that encodes genes whose functions are typically restricted to the mitochondrion.

In some cases, the number of mitochondria per cell is tunable, depending, typically, on energy demand. It is for instance possible muscle cells that are used—that by extension have a higher demand for ATP—may often be found to have a significantly higher number of mitochondria than cells that do not have a high energy load.

The structure of the mitochondria can vary significantly depending on the organism and the state of the cell cycle which one is observing. The typical textbook image, however, depicts mitochondria as oval-shaped organelles with a double inner and outer membrane ; learn to recognize this generic representation. Both the inner and outer membranes are phospholipid bilayers embedded with proteins that mediate transport across them and catalyze various other biochemical reactions. The inner membrane layer has folds called cristae that increase the surface area into which respiratory chain proteins can be embedded. The region within the cristae is called the mitochondrial matrix and contains—among other things—enzymes of the TCA cycle. During respiration, protons are pumped by respiratory chain complexes from the matrix into a region known as the intermembrane space (between the inner and outer membranes).

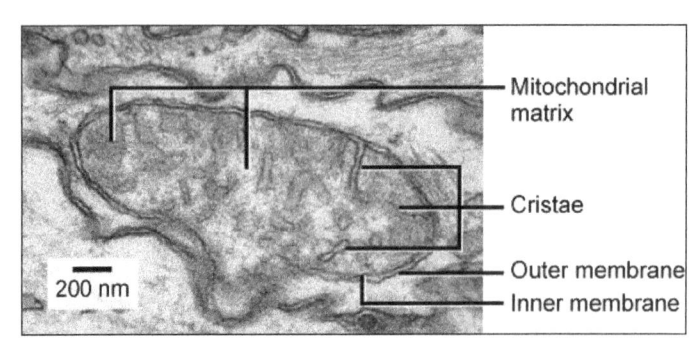

This electron micrograph shows a mitochondrion as viewed with a transmission electron microscope. This organelle has an outer membrane and an inner membrane. The inner membrane contains folds, called cristae, which increase its surface area. The space between the two membranes is called the intermembrane space, and the space inside the inner membrane is called the mitochondrial matrix. ATP synthesis takes place on the inner membrane.

Peroxisomes

Peroxisomes are small, round organelles enclosed by single membranes. These organelles carry out redox reactions that oxidize and break down fatty acids and amino acids. They also help to detoxify many toxins that may enter the body. Many of these redox reactions release hydrogen peroxide, H_2O_2, which would be damaging to cells; however, when these reactions are confined to peroxisomes, enzymes safely break down the H_2O_2 into oxygen and water. For example, alcohol is detoxified by peroxisomes in liver cells. Glyoxysomes, which are specialized peroxisomes in plants, are responsible for converting stored fats into sugars.

Vesicles and Vacuoles

Vesicles and vacuoles are membrane-bound sacs that function in storage and transport. Other than the fact that vacuoles are somewhat larger than vesicles, there is a very subtle distinction between them: the membranes of vesicles can fuse with either the plasma membrane or other membrane systems within the cell. Additionally, some agents such as enzymes within plant vacuoles break down macromolecules. The membrane of a vacuole does not fuse with the membranes of other cellular components.

Animal Cells versus Plant Cells

At this point, you know that each eukaryotic cell has a plasma membrane, cytoplasm, a nucleus, ribosomes, mitochondria, peroxisomes, and in some, vacuoles. There are some striking differences between animal and plant cells worth noting. Here is a brief list of differences that we want you to be familiar with and a slightly expanded description below:

1. While all eukaryotic cells use microtubule and motor protein the based mechanisms to segregate chromosomes during cell division, the structures used to organize these microtubules

differ in plants versus animal and yeast cells. Animal and yeast cells organize and anchor their microtubules into structures called microtubule organizing centers (MTOCs). These structures are composed of structures called centrioles that are composed largely of α-tubulin, β-tubulin, and other proteins. Two centrioles organize into a structure called a centrosome. By contrast, in plants, while microtubules also organize into discrete bundles, there are no conspicuous structures similar to the MTOCs seen in animal and yeast cells. Rather, depending on the organism, it appears that there can be several places where these bundles of microtubules can nucleate from places called acentriolar (without centriole) microtubule organizing centers. A third type of tubulin, γ-tubulin, appears to be implicated, but our knowledge of the precise mechanisms used by plants to organize microtubule spindles is still spotty.

2. Animal cells typically have organelles called lysosomes responsible for degradation of biomolecules. Some plant cells contain functionally similar degradative organelles, but there is a debate as to how they should be named. Some plant biologists call these organelles lysosomes while others lump them into the general category of plastids and do not give them a specific name.

3. Plant cells have a cell wall, chloroplasts and other specialized plastids, and a large central vacuole, whereas animal cells do not.

The Centrosome

The centrosome is a microtubule-organizing center found near the nuclei of animal cells. It contains a pair of centrioles, two structures that lie perpendicular to eachother . Each centriole is a cylinder of nine triplets of microtubules.

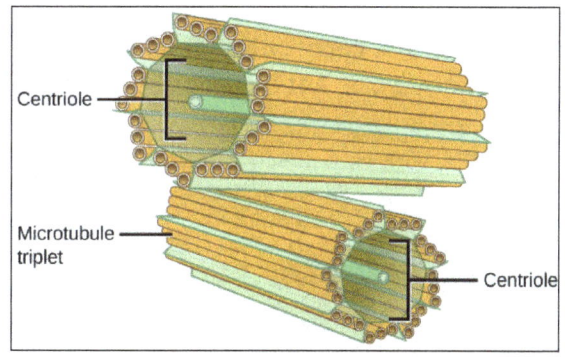

Figure: The centrosome consists of two centrioles that lie at right angles to each other. Each centriole is a cylinder made up of nine triplets of microtubules. Nontubulin proteins (indicated by the green lines) hold the microtubule triplets together.

The centrosome (the organelle where all microtubules originate in animal and yeast) replicates itself before a cell divides, and the centrioles appear to have some role in pulling the duplicated chromosomes to opposite ends of the dividing cell. However, the exact function of the centrioles in cell division remains unclear, as cells that have had their centrosome removed can still divide, and plant cells, which lack centrosomes, are capable of cell division.

Lysosomes

Animal cells have another set of organelles not found in plant cells: lysosomes. Colloquially, the lysosomes are sometimes called the cell's "garbage disposal". Enzymes within the lysosomes aid the breakdown of proteins, polysaccharides, lipids, nucleic acids, and even "worn-out" organelles. These enzymes are active at a much lower pH than that of the cytoplasm. Therefore, the pH within lysosomes is more acidic than the pH of the cytoplasm. In plant cells, many of the same digestive processes take place in vacuoles.

The Cell Wall

If you examine the diagram above depicting plant and animal cells, you will see in the diagram of a plant cell a structure external to the plasma membrane called the cell wall. The cell wall is a rigid covering that protects the cell, provides structural support, and gives shape to the cell. Fungal and protistan cells also have cell walls. While the chief component of bacterial cell walls is peptidoglycan, the major organic molecule in the plant cell wall is cellulose, a polysaccharide made up of glucose subunits.

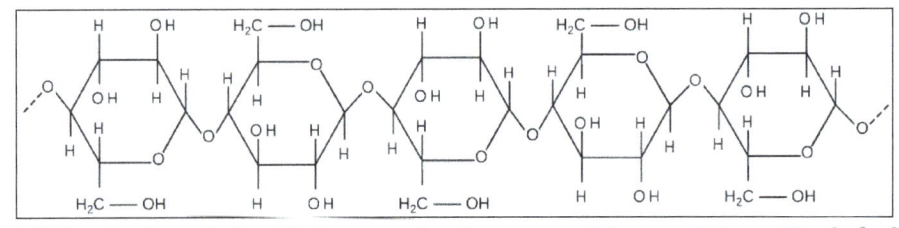

Figure: Cellulose is a long chain of β-glucose molecules connected by a 1-4 linkage. The dashed lines at each end of the figure indicate a series of many more glucose units. The size of the page makes it impossible to portray an entire cellulose molecule.

Chloroplasts

Chloroplasts are plant cell organelles that carry out photosynthesis. Like the mitochondria, chloroplasts have their own DNA and ribosomes, but chloroplasts have an entirely different function.

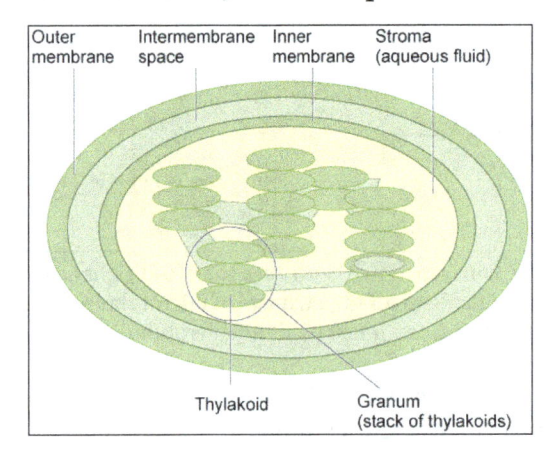

The chloroplast has an outer membrane, an inner membrane, and membrane structures called thylakoids that are stacked into grana. The space inside the thylakoid membranes is called the thylakoid space. The light harvesting reactions take place in the thylakoid membranes, and the synthesis of sugar takes place in the fluid inside the inner membrane, which is called the stroma. Chloroplasts also have their own genome, which is contained on a single circular chromosome.

Like mitochondria, chloroplasts have outer and inner membranes, but within the space enclosed by a chloroplast's inner membrane is a set of interconnected and stacked fluid-filled membrane sacs called thylakoids (figure below). Each stack of thylakoids is called a granum (plural = grana). The fluid enclosed by the inner membrane that surrounds the grana is called the stroma.

The chloroplasts contain a green pigment called chlorophyll, which captures the light energy that drives the reactions of photosynthesis. Like plant cells, photosynthetic protists also have chloroplasts. Some bacteria perform photosynthesis, but their chlorophyll is not relegated to an organelle.

The Central Vacuole

Previously, we mentioned vacuoles as essential components of plant cells. The central vacuole plays a key role in regulating the cell's concentration of water in changing environmental conditions.

Silly vacuole factoid: Have you ever noticed that if you forget to water a plant for a few days, it wilts? That's because as the water concentration in the soil becomes lower than the water concentration in the plant, water moves out of the central vacuoles and cytoplasm. As the central vacuole shrinks, it leaves the cell wall unsupported. This loss of support to the cell walls of plant cells results in the wilted appearance of the plant.

The central vacuole also supports the expansion of the cell. When the central vacuole holds more water, the cell gets larger without having to invest a lot of energy in synthesizing new cytoplasm.

Prokaryotic Cell

Prokaryotic cells are cells that do not have a true nucleus or membrane-bound organelles. Organisms within the domains Bacteria and Archaea have prokaryotic cells, while other forms of life are eukaryotic. However, organisms with prokaryotic cells are abundant and make up much of Earth's biomass.

Organisms that have prokaryotic cells are unicellular and are called prokaryotes. Prokaryotic cells can be contrasted with eukaryotic cells, which are more complex. Eukaryotic cells have a nucleus surrounded by a nuclear membrane and also have other organelles that perform specific functions in the cell. A prokaryotic cell contains only a single membrane, which surrounds the cell as an outer membrane.

All of the reactions within a prokaryote, therefore, take place within the cytoplasm of the cell. While this makes the cells slightly less efficient, prokaryotic cells still have a remarkable reproductive capacity. A prokaryote reproduces through binary fission, a process which simply splits duplicated DNA into separate cells. Without any organelles or complex chromosomes to reproduce, most prokaryotic cells can divide every 24 hours, or even faster with an adequate supply of food.

While many prokaryotic cells have adapted to free-living within the environment, many more have adapted to live within the gut of other organisms. These commensal organisms survive by breaking down molecules inside the gut and allow the organism they are living within the ability to digest a wider variety of foods. For example, the human gut contains 2-3 pounds of bacteria, which have evolved to help us digest complex carbohydrates, proteins, and fats.

Examples of Prokaryotic Cells

Bacterial Cells

Bacteria are single-celled microorganisms that are found nearly everywhere on Earth, and they are very diverse in their shapes and structures. There are about 5×10^{30} bacteria living on Earth, including in our own bodies; in the human gut, bacteria outnumber human cells 10:1.

The cell walls of bacteria contain peptidoglycan, a molecule made of sugars and amino acids that gives the cell wall its structure and is thicker in some bacteria than others. Bacteria contain certain structures unique to them as previously mentioned, such as the capsule, flagella, and pili. Most bacteria have just one chromosome that is circular, which can range from about 160,000 base pairs (bp) to 12,200,000 bp. They also contain plasmids, which are small circular pieces of DNA that replicate independently of the chromosome.

Some bacteria can form endospores. These are tough, dormant structures that the bacteria can reduce themselves to under starvation conditions when not enough nutrients are available. They do not need nutrients and are resistant to extreme temperatures, UV rays, and chemicals. When environmental conditions become favorable again, the endospore can reactivate.

Archaeal Cells

Archaea are similar in size and shape to bacteria, and they are also unicellular. Since bacteria and archaea are the two types of prokaryotes, this means that all prokaryotes are unicellular. Some archaea are found in extreme environments, such as hot springs, but they can be found in a variety of locations, such as soils, oceans, marshlands, and inside other organisms, including humans.

Like bacteria, archaea can have a cell wall and flagella. However, the structure of these organelles is different. For example, archaeal cell walls do not contain peptidoglycan. In addition, the flagella of archaea work the same way as those of bacteria, but they evolved from different structures. Membranes of archaea are very different than those of all other lifeforms; they contain different lipids, which have a different stereochemistry. Archaea usually have one circular chromosome, as bacteria do. The archaeal chromosome can range from less than 491,000 bp to about 5,700,000 bp. They can also contain plasmids.

Prokaryotic Cell Structure

Prokaryotic cells do not have a true nucleus that contains their genetic material as eukaryotic cells do. Instead, prokaryotic cells have a nucleoid region, which is an irregularly-shaped region that contains the cell's DNA and is not surrounded by a nuclear envelope. Some other parts of prokaryotic cells are similar to those in eukaryotic cells, such as a cell wall surrounding the cell (which is also found in plant cells, although it has a different composition).

Like eukaryotic cells, prokaryotic cells have cytoplasm, a gel-like substance that makes up the "filling" of the cell, and a cytoskeleton that holds components of the cell in place. Both prokaryotic cells and eukaryotic cells have ribosomes, which are organelles that produce proteins, and vacuoles, small spaces in cells that store nutrients and help eliminate waste.

Some prokaryotic cells have flagella, which are tail-like structures that enable the organism to move around. They may also have pili, small hair-like structures that help bacteria adhere to surfaces and can allow DNA to be transferred between two prokaryotic cells in a process known as conjugation. Another part that is found in some bacteria is the capsule. The capsule is a sticky layer of carbohydrates that helps the bacterium adhere to surfaces in its surroundings.

Prokaryotic Cell Diagram

The following image is a diagram of a prokaryotic cell; in this case, a bacterium.

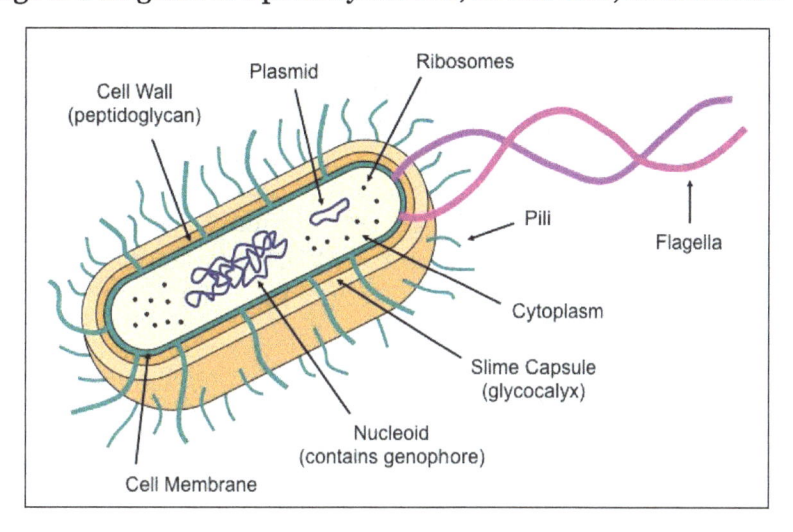

Characteristics of Prokaryotic Cells

All prokaryotic cells have a nucleoid region, DNA and RNA as their genetic material, ribosomes that make proteins, and cytoplasm that contains a cytoskeleton, which organizes and supports the parts of the cell. Prokaryotic cells are simpler than eukaryotic cells, and an organism that is a prokaryote is unicellular; it is made up of only one prokaryotic cell.

Prokaryotic cells are usually between 0.1 to 5 micrometers in length (.00001 to .0005 cm). Eukaryotic cells are generally much larger, between 10 and 100 micrometers. Prokaryotic cells have a higher surface-area-to-volume ratio because they are smaller, which makes them able to obtain a larger amount of nutrients via their plasma membrane.

Prokaryotic Cell Parts

Unlike eukaryotic cells, prokaryotic cells have no distinct organelles bound by membranes. Instead, the many reactions the cell conducts happen within the cytoplasm of the cell. In fact, there are 2 main components that are present within all prokaryotic cells.

The first is a cell membrane. This is a layer of phospholipid molecules which separate the inside of the cell from the outside. While not present in all prokaryotes, many secrete a cell wall, used to protect and house the cell in an extra layer of proteins and structural molecules.

The second part found in all prokaryotic cells is DNA. DNA is the basic blueprint for all life and is found within all cells. In prokaryotes, the DNA often takes the form of a large circular genome.

This can be compared to the organized chromosomes which are typically found within eukaryotes. This large circle of DNA directs which proteins the cell creates, and regulates the actions of the cell.

Other prokaryotic cells can have a large number of different parts, such as cilia and flagella to help them move around. While these structures are similar in function to those found within eukaryotes, they often have a different structure. This suggests that the two types of cell have undergone very different selection processes and have independently involved the structures.

Prokaryotic Cells Division

Prokaryotic cells divide through the process of binary fission. Unlike mitosis, this process does not involve the condensation of DNA or the duplication of organelles. Prokaryotic cells have only a small amount of DNA, which is not stored in complex chromosomes. Further, there are no organelles so there is nothing to divide.

When a prokaryote grows to a large size, the process of binary fission takes place. This process duplicates the DNA, then separates each new strand of DNA into individual cells. This process is simpler than mitosis, and as such bacteria can reproduce much faster.

Cell Functions

A cell performs six major functions essential for the growth and development of an organism. The functions of a cell include:

Provides Support and Structure

All the organisms are made up of cells. They form the structural basis of all the organisms. The cell wall and the cell membrane are the main components that function to provide support and structure to the organism. For eg., the skin is made up of a large number of cells. Xylem present in the vascular plants is made of cells that provide structural support to the plants.

Facilitate Growth Mitosis

In the process of mitosis, the parent cell divides into the daughter cells. Thus, the cells multiply and facilitate the growth in an organism.

Allows Transport of Substances

Various nutrients are imported by the cells to carry out various chemical processes going on inside the cells. The waste produced by the chemical processes is eliminated from the cells by active and passive transport.

Small molecules such as oxygen, carbon dioxide, and ethanol diffuse across the cell membrane along the concentration gradient. This is known as passive transport.

The larger molecules diffuse across the cell membrane through active transport where the cells require a lot of energy to transport the substances.

Cell: The Fundamental Unit of Life

Energy Production

Cells require energy to carry out various chemical processes. This energy is produced by the cells through a process called photosynthesis in plants and respiration in animals.

Aids in Reproduction

A cell aids in reproduction through the processes called mitosis and meiosis. Mitosis is termed as the asexual reproduction where the parent cell divides to form daughter cells. Meiosis causes the daughter cells to be genetically different from the parent cells.

Thus, we see, why cells are known as the structural and functional unit of life. This is because they are responsible for providing structure to the organisms and performs several functions necessary for carrying out life's processes.

Cellular Respiration

Cellular respiration is the process by which living cells break down glucose molecules and release energy. The process is similar to burning, although it doesn't produce light or intense heat as a campfire does. This is because cellular respiration releases the energy in glucose slowly, in many small steps. It uses the energy that is released to form molecules of ATP, the energy-carrying molecules that cells use to power biochemical processes. Cellular respiration involves many chemical reactions, but they can all be summed up with this chemical equation:

$$C_6H_{12}O_6 + 6O_2 \rightarrow 6CO_2 + 6H_2O + \text{Chemical Energy (In ATP)}$$

In words, the equation shows that glucose ($C_6H_{12}O_6$) and oxygen (O_2) react to form carbon dioxide (CO_2 and water H_2O, releasing energy in the process. Because oxygen is required for cellular respiration, it is an aerobic process.

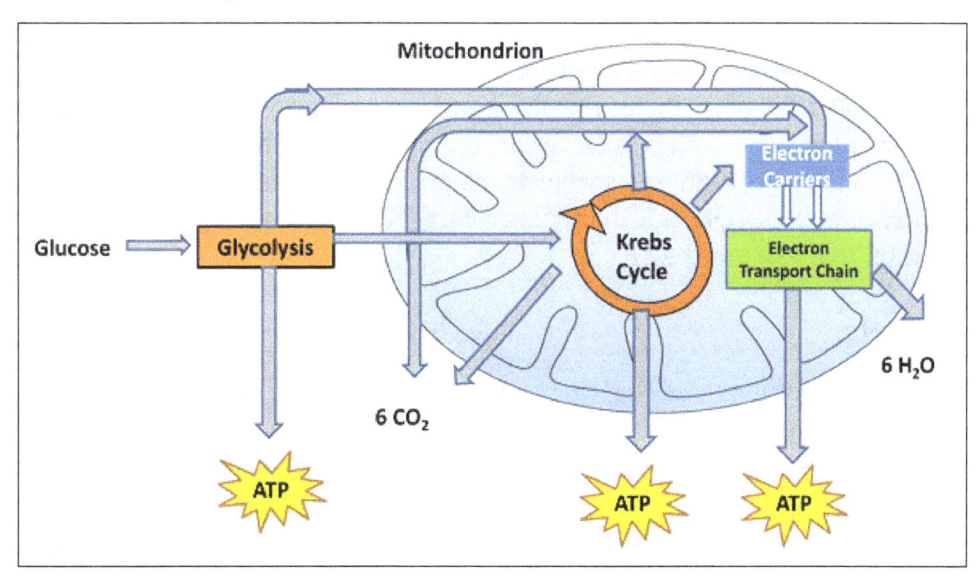

Cellular respiration takes place in the stages shown here. The process begins with a molecule of glucose, which has six carbon atoms.

Cellular respiration occurs in the cells of all living things, both autotrophs and heterotrophs. All of them burn glucose to form ATP. The reactions of cellular respiration can be grouped into three stages: glycolysis, the Krebs cycle (also called the citric acid cycle), and electron transport. The figure above gives an overview of these three stages, which are also described in detail below.

Glycolysis

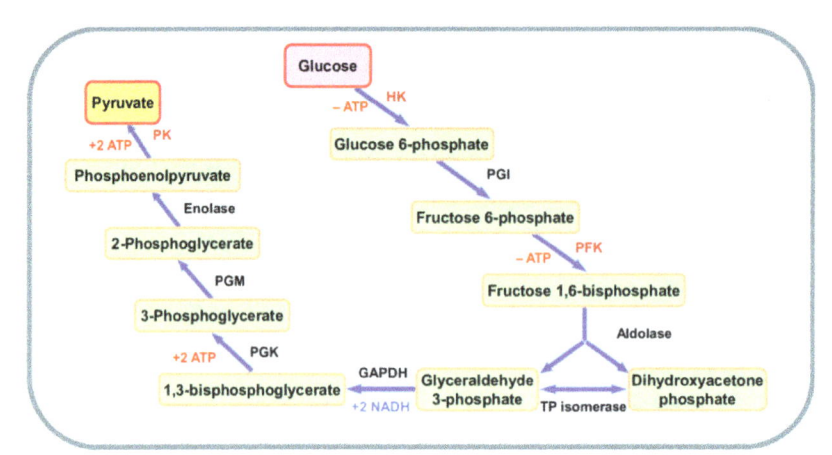

Figure: In glycolysis, a glucose molecule is converted into two pyruvate molecules.

The first stage of cellular respiration is glycolysis. It takes place in the cytosol of the cytoplasm.

Splitting Glucose

The word glycolysis means "glucose splitting," which is exactly what happens in this stage. Enzymes split a molecule of glucose into two molecules of pyruvate (also known as pyruvic acid). This occurs in several steps, as shown in the following diagram.

Results of Glycolysis

Energy is needed at the start of glycolysis to split the glucose molecule into two pyruvate molecules. These two molecules go on to stage II of cellular respiration. The energy to split glucose is provided by two molecules of ATP. As glycolysis proceeds, energy is released, and the energy is used to make four molecules of ATP. As a result, there is a net gain of two ATP molecules during glycolysis. During this stage, high-energy electrons are also transferred to molecules of NAD+ to produce two molecules of NADH, another energy-carrying molecule. NADH is used in stage III of cellular respiration to make more ATP.

Citric Acid Cycle

In eukaryotic cells, the pyruvate molecules produced at the end of glycolysis are transported into mitochondria, which are sites of cellular respiration. If oxygen is available, aerobic respiration will go forward. In mitochondria, pyruvate will be transformed into a two-carbon acetyl group (by removing a molecule of carbon dioxide) that will be picked up by a carrier compound called coenzyme A (CoA), which is made from vitamin B5. The resulting compound is called acetyl CoA. Acetyl CoA can be used in a variety of ways by the cell, but its major function is to deliver the acetyl group derived from pyruvate to the next pathway in glucose catabolism.

Cell: The Fundamental Unit of Life

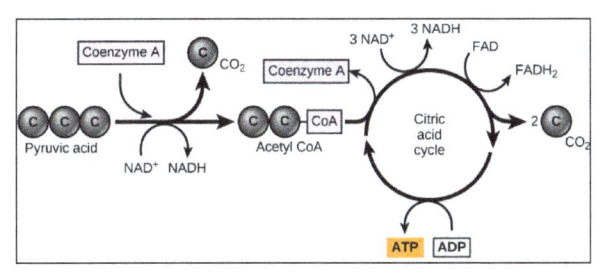

Figure: Pyruvate is converted into acetyl-CoA before entering the citric acid cycle.

Structure of the Mitochondrion

Before you read about the last two stages of cellular respiration, you need to review the structure of the mitochondrion, where these two stages take place.

As you can see from the figure, a mitochondrion has an inner and outer membrane. The space between the inner and outer membrane is called the intermembrane space. The space enclosed by the inner membrane is called the matrix. The second stage of cellular respiration, the Krebs cycle, takes place in the matrix. The third stage, electron transport, takes place on the inner membrane.

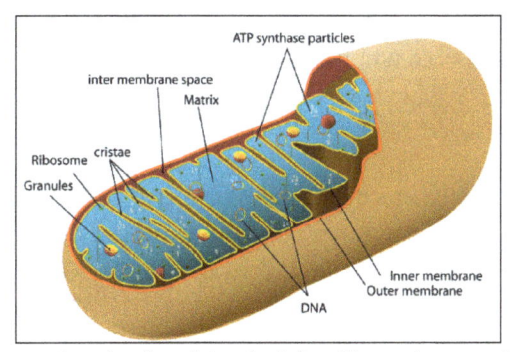

Figure: The structure of a mitochondrion is defined by an inner and outer membrane.

Krebs Cycle

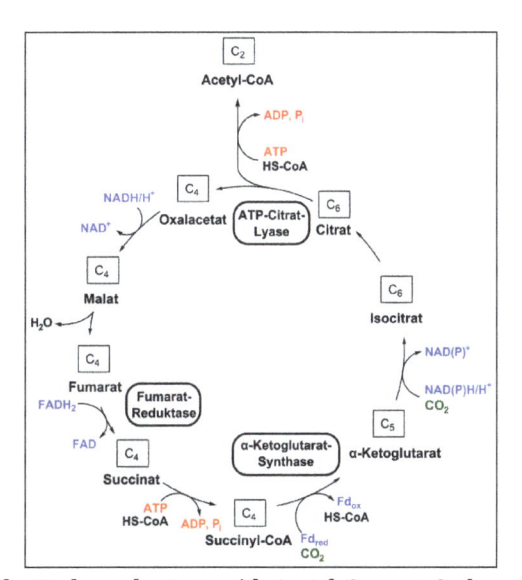

Figure: The Krebs cycle starts with Acetyl-Coa, a 2 Carbon molecule.

Recall that glycolysis produces two molecules of pyruvate (pyruvic acid). These molecules enter the matrix of a mitochondrion, where they start the Krebs cycle. The reactions that occur next are shown in the following figure.

Before the Krebs cycle begins, pyruvic acid, which has three carbon atoms, is split apart and combined with an enzyme known as CoA, which stands for coenzyme A. The product of this reaction is a two-carbon molecule called acetyl-CoA. The third carbon from pyruvic acid combines with oxygen to form carbon dioxide, which is released as a waste product. High-energy electrons are also released and captured in NADH.

Steps of the Krebs Cycle

The Krebs cycle itself actually begins when acetyl-CoA combines with a four-carbon molecule called OAA (oxaloacetate). This produces citric acid, which has six carbon atoms. This is why the Krebs cycle is also called the citric acid cycle. After citric acid forms, it goes through a series of reactions that release energy. The energy is captured in molecules of NADH, ATP, and $FADH_2$ (another energy-carrying compound.) Carbon dioxide is also released as a waste product of these reactions. The final step of the Krebs cycle regenerates OAA, the molecule that began the Krebs cycle. This molecule is needed for the next turn through the cycle. Two turns are needed because glycolysis produces two pyruvic acid molecules when it splits glucose.

Results of the Krebs Cycle

After the second turn through the Krebs cycle, the original glucose molecule has been broken down completely. All six of its carbon atoms have combined with oxygen to form carbon dioxide. The energy from its chemical bonds has been stored in a total of 16 energy-carrier molecules. These molecules are:

- 4 ATP (including 2 from glycolysis).
- 10 NADH (including 2 from glycolysis).
- 2 FADH22.

Oxidative Phosphorylation

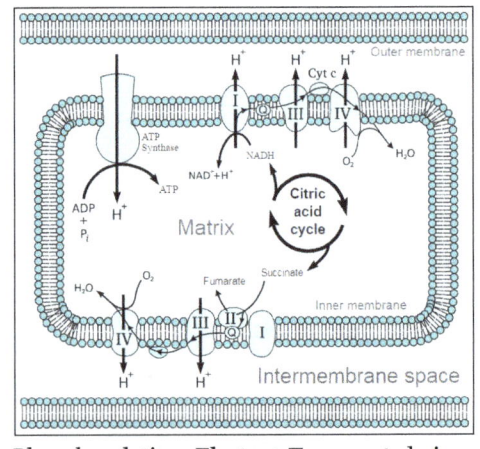

Figure: Oxidative Phosphorylation: Electron Transport chain and Chemiosmosis.

Oxidative phosphorylation is the final stage of aerobic cellular respiration. There are two substages of oxidative phosphorylation, Electron transport chain and Chemiosmosis. In these stages, energy from NADH and $FADH_2$, which result from the previous stages of the cellular respiration, is transferred to ATP.

Electron Transport Chain

During this stage, high-energy electrons are released from NADH and FADH2, and they move along electron-transport chains on the inner membrane of the mitochondrion. An electron-transport chain is a series of molecules that transfer electrons from molecule to molecule by chemical reactions. Some of the energy from the electrons is used to pump hydrogen ions (H+) across the inner membrane, from the matrix into the intermembrane space. This ion transfer creates an electrochemical gradient that drives the synthesis of ATP.

Chemiosmosis

The pumping of hydrogen ions across the inner membrane creates a greater concentration of the ions in the intermembrane space than in the matrix. This gradient causes the ions to flow back across the membrane into the matrix, where their concentration is lower. the ATP synthase acts as a channel protein, helping the hydrogen ions across the membrane. It also acts as an enzyme, forming ATP from ADP and inorganic phosphate in a process called oxidative phosphorylation. After passing through the electron-transport chain, the "spent" electrons combine with oxygen to form water.

Cellular Reproduction

One of the most defining characteristics of the living condition is the ability to reproduce. All living cells and organisms reproduce, producing offspring like themselves, and pass on the hereditary information contained in their DNA molecules. The processes of reproduction, while varied and complex, depends upon the ability of individual cells to replicate. All cells arise from preexisting cells by some mechanism of cell reproduction or division.

Cell reproduction is often divided into two major types: asexual and sexual cell reproduction. Typically in asexual reproduction, a single cell gives rise to a genetic duplicate of the parental cell, without any genetic contribution from another individual, while sexual cell reproduction involves the genetic recombination of two cells.

In procaryotes, asexual cell division often proceeds by a process of fission. Reproduction occurs when a parental cell replicates its bacterial DNA producing a complete and faithful copy of its chromosome. Growth of the bacterial cell to an appropriate size seems to induce division by binary fission. New plasma membranes and wall material are laid down constricting the cell into two pieces. The splitting of the cytoplasmic domains and the two DNA molecules into nearly equal halves results in two daughter cells.

Budding, which is another method of asexual reproduction, occurs in most yeast, hydra, and in some filamentous fungi. In this process, a small cytoplasmic swelling protrudes (a bud) develops on the surface of either the yeast cell or the hypha, with its cytoplasm being continuous with that of the parent cell. The parental cell nucleus then divides and one of the daughter nuclei migrates into the bud. Continuous synthesis of cytoplasm and repeated nuclear divisions results in many buds over the cells surface. Buds are pinched off and behave as spores, germinating and forming a new hypha, all genetically identical to the parent.

In higher organisms (most eucaryotes) asexual cell reproduction involves an elaborate duplication of the chromosomes followed by their separation in a nuclear division called mitosis. Mitosis is often followed by cytokinesis, a division of the cytoplasm. In the hard-walled cells of higher plants, a medial cell plate forms and divides the parental cell into two compartments. In animal cells, which do not have a hard cell wall, a membrane furrow, made of a microfilament contractile ring, that constricts as a camera diaphragm, pinching the cell in two daughter cells.

Cells that reproduce sexually are characterized by meiosis, the nuclear division process by which sex cells (gametes) are formed. Every chromosome of a somatic cell occurs in a pair (diploid). During meiosis these diploid pairs of chromosomes duplicate and are separated so that each meiotic sex cell has only one chromosome (haploid) of each pair. Two successive meiotic divisions result in the production of haploid sperm and egg cells, each with one-half of the amount of the parental DNA.

During the life cycle of sexually reproducing organisms, fertilization results in the fusion of haploid gametes (sperm and egg) producing the zygote. Dividing by asexual cell reproduction, the zygote undergoes cellular differentiation, whereby cells become structurally, functionally, and biochemically distinct from each other.

Cell Cycle

The cell cycle can be thought of as the life cycle of a cell. In other words, it is the series of growth and development steps a cell undergoes between its "birth"—formation by the division of a mother cell—and reproduction—division to make two new daughter cells.

Stages of the Cell Cycle

To divide, a cell must complete several important tasks: it must grow, copy its genetic material (DNA), and physically split into two daughter cells. Cells perform these tasks in an organized, predictable series of steps that make up the cell cycle. The cell cycle is a cycle, rather than a linear pathway, because at the end of each go-round, the two daughter cells can start the exact same process over again from the beginning.

In eukaryotic cells, or cells with a nucleus, the stages of the cell cycle are divided into two major phases: interphase and the mitotic (M) phase.

- During interphase, the cell grows and makes a copy of its DNA.

- During the mitotic (M) phase, the cell separates its DNA into two sets and divides its cytoplasm, forming two new cells.

Interphase

Let's enter the cell cycle just as a cell forms, by division of its mother cell. What must this newborn cell do next if it wants to go on and divide itself? Preparation for division happens in three steps:

- G_1. During G_1, also called the first gap phase, the cell grows physically larger, copies organelles, and makes the molecular building blocks it will need in later steps.
- S phase. In S phase, the cell synthesizes a complete copy of the DNA in its nucleus. It also duplicates a microtubule-organizing structure called the centrosome. The centrosomes help separate DNA during M phase.
- G_2 phase. During the second gap phase, or G_2 phase, the cell grows more, makes proteins and organelles, and begins to reorganize its contents in preparation for mitosis. G_2 phase ends when mitosis begins.

The G_1, S, and G_2, phases together are known as interphase. The prefix inter- means between, reflecting that interphase takes place between one mitotic (M) phase and the next.

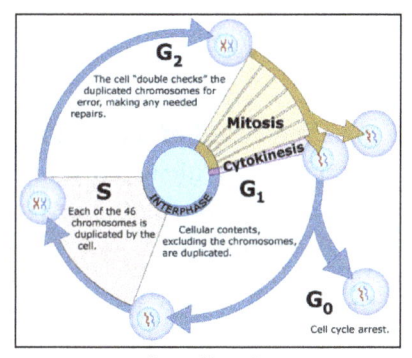

The cell cycle.

M Phase

During the mitotic (M) phase, the cell divides its copied DNA and cytoplasm to make two new cells. M phase involves two distinct division-related processes: mitosis and cytokinesis.

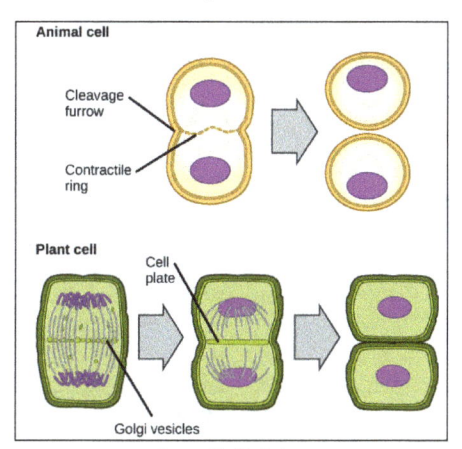

The cell division.

In mitosis, the nuclear DNA of the cell condenses into visible chromosomes and is pulled apart by the mitotic spindle, a specialized structure made out of microtubules. Mitosis takes place in four stages: prophase (sometimes divided into early prophase and prometaphase), metaphase, anaphase, and telophase.

In cytokinesis, the cytoplasm of the cell is split in two, making two new cells. Cytokinesis usually begins just as mitosis is ending, with a little overlap. Importantly, cytokinesis takes place differently in animal and plant cells.

- In animals, cell division occurs when a band of cytoskeletal fibers called the contractile ring contracts inward and pinches the cell in two, a process called contractile cytokinesis. The indentation produced as the ring contracts inward is called the cleavage furrow. Animal cells can be pinched in two because they're relatively soft and squishy.

- Plant cells are much stiffer than animal cells; they're surrounded by a rigid cell wall and have high internal pressure. Because of this, plant cells divide in two by building a new structure down the middle of the cell. This structure, known as the cell plate, is made up of plasma membrane and cell wall components delivered in vesicles, and it partitions the cell in two.

Cell cycle exit and G_0

What happens to the two daughter cells produced in one round of the cell cycle? This depends on what type of cells they are. Some types of cells divide rapidly, and in these cases, the daughter cells may immediately undergo another round of cell division. For instance, many cell types in an early embryo divide rapidly, and so do cells in a tumor.

Other types of cells divide slowly or not at all. These cells may exit the G_1 phase and enter a resting state called G_0 phase. In G_0, a cell is not actively preparing to divide, it's just doing its job. For instance, it might conduct signals as a neuron (like the one in the drawing below) or store carbohydrates as a liver cell. G_0 is a permanent state for some cells, while others may re-start division if they get the right signals.

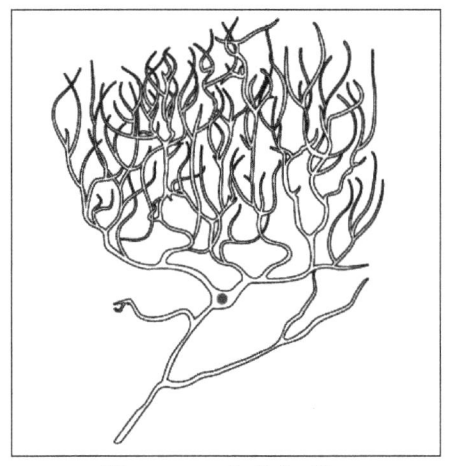

Neurons and glial cells.

Duration of the Cell Cycle

Different cells take different lengths of time to complete the cell cycle. A typical human cell might take about 24 hours to divide, but fast-cycling mammalian cells, like the ones that line the intestine, can complete a cycle every 9-10 hours when they're grown in culture.

Different types of cells also split their time between cell cycle phases in different ways. In early frog embryos, for example, cells spend almost no time in G_1 and G_2 and instead rapidly cycle between S and M phases—resulting in the division of one big cell, the zygote, into many smaller cells.

References

- Cell-biology, science: britannica.com, Retrieved 1 May, 2019
- Cell-theory: biologydictionary.net, Retrieved 15 January, 2019
- Structure-and-Function, Eukaryotic-Cell, Introductory-Biology: libretexts.org, Retrieved 14 June, 2019
- Prokaryotic-cell: biologydictionary.net, Retrieved 23 March, 2019
- Cells, biology: byjus.com, Retrieved 3 August, 2019
- Cellular-Respiration, Cells, Human-Biology: libretexts.org, Retrieved 21 June, 2019
- Reproduction: cox.miami.edu, Retrieved 24 April, 2019
- Cell-cycle-phases, cellular-molecular-biology, biology, science: khanacademy.org, Retrieved 12 February, 2019

Chapter 3
Reproduction in Organisms

The biological process by which new individual organisms are produced from their parents is known as reproduction. Sexual and asexual are the two forms of reproduction. Plant reproduction consists of both sexual and asexual reproduction whereas human reproduction is a form of sexual reproduction. The topics elaborated in this chapter will help in gaining a better perspective about these types of reproduction as well as the processes related to them.

Reproduction

Reproduction is the process by which organisms replicate themselves. In a general sense reproduction is one of the most important concepts in biology: it means making a copy, a likeness, and thereby providing for the continued existence of species. Although reproduction is often considered solely in terms of the production of offspring in animals and plants, the more general meaning has far greater significance to living organisms. To appreciate this fact, the origin of life and the evolution of organisms must be considered. One of the first characteristics of life that emerged in primeval times must have been the ability of some primitive chemical system to make copies of itself.

At its lowest level, therefore, reproduction is chemical replication. As evolution progressed, cells of successively higher levels of complexity must have arisen, and it was absolutely essential that they had the ability to make likenesses of themselves. In unicellular organisms, the ability of one cell to reproduce itself means the reproduction of a new individual; in multicellular organisms, however, it means growth and regeneration. Multicellular organisms also reproduce in the strict sense of the term—that is, they make copies of themselves in the form of offspring—but they do so in a variety of ways, many involving complex organs and elaborate hormonal mechanisms.

Levels of Reproduction

Molecular Replication

The characteristics that an organism inherits are largely stored in cells as genetic information in very long molecules of deoxyribonucleic acid (DNA). In 1953 it was established that DNA molecules consist of two complementary strands, each of which can make copies of the other. The strands are like two sides of a ladder that has been twisted along its length in the shape of a double helix (spring). The rungs, which join the two sides of the ladder, are made up of two terminal bases. There are four bases in DNA: thymine, cytosine, adenine, and guanine. In the middle of each rung a base from one strand of DNA is linked by a hydrogen bond to a base of the other strand. But they can pair only in certain ways: adenine always pairs with thymine, and guanine with cytosine. This is why one strand of DNA is considered complementary to the other.

The double helices duplicate themselves by separating at one place between the two strands and becoming progressively unattached. As one strand separates from the other, each acquires new complementary bases until eventually each strand becomes a new double helix with a new complementary strand to replace the original one. Because adenine always falls in place opposite thymine and guanine opposite cytosine, the process is called a template replication—one strand serves as the mold for the other. It should be added that the steps involving the duplication of DNA do not occur spontaneously; they require catalysts in the form of enzymes that promote the replication process.

Molecular Reproduction

The sequence of bases in a DNA molecule serves as a code by which genetic information is stored. Using this code, the DNA synthesizes one strand of ribonucleic acid (RNA), a substance that is so similar structurally to DNA that it is also formed by template replication of DNA. RNA serves as a messenger for carrying the genetic code to those places in the cell where proteins are manufactured. The way in which the messenger RNA is translated into specific proteins is a remarkable and complex process. The ability to synthesize enzymes and other proteins enables the organism to make any substance that existed in a previous generation. Proteins are reproduced directly; however, such other substances as carbohydrates, fats, and other organic molecules found in cells are produced by a series of enzyme-controlled chemical reactions, each enzyme being derived originally from DNA through messenger RNA. It is because all of the organic constituents made by organisms are derived ultimately from DNA that molecules in organisms are reproduced exactly by each successive generation.

Cell Reproduction

The chemical constituents of cytoplasm (that part of the cell outside the nucleus) are not resynthesized from DNA every time a cell divides. This is because each of the two daughter cells formed during cell division usually inherits about half of the cellular material from the mother cell, and is important because the presence of essential enzymes enables DNA to replicate even before it has made the enzymes necessary to do so.

Cells of higher organisms contain complex structures, and each time a cell divides the structures must be duplicated. The method of duplication varies for each structure, and in some cases the mechanism is still uncertain. One striking and important phenomenon is the formation of a new membrane. Cell membranes, although they are very thin and appear to have a simple form and structure, contain many enzymes and are sites of great metabolic activity. This applies not only to the membrane that surrounds the cell but to all the membranes within the cell. New membranes, which seem to form rapidly, are indistinguishable from old ones.

Thus, the formation of a new cell involves the further synthesis of many constituents that were present in the parent cell. This means that all of the information and materials necessary for a cell to reproduce itself must be supplied by the cellular constituents and the DNA inherited from the parent cell.

Binary Fission

Of the various kinds of cell division, the most common mode is binary fission, the division of a cell into two separate and similar parts. In bacteria (prokaryotes) the chromosome (the body that contains the DNA and associated proteins) replicates and then divides in two, after which a cell wall forms across the elongated parent cell. In higher organisms (eukaryotes) there is first an elaborate duplication and then a separation of the chromosomes (mitosis), after which the cytoplasm divides in two. In the hard-walled cells of higher plants, a median plate forms and divides the mother cell into two compartments; in animal cells, which do not have a hard wall, a delicate membrane pinches the cell in two, much like the separation of two liquid drops. Budding yeast cells provide an interesting exception. In these fungi the cell wall forms a bubble that becomes engorged with cytoplasm until it is ultimately the size of the original cell. The nucleus then divides, one of the daughter nuclei passes into the bud, and ultimately the two cells separate.

In some instances of binary fission, there may be an unequal cytoplasmic division with an equal division of the chromosomes. This occurs, in fact, in a large number of higher organisms during meiosis—the process by which sex cells (gametes) are formed: originally each chromosome of the cell is in a pair (diploid); during meiosis these diploid pairs of chromosomes are separated so that each sex cell has only one of each pair of chromosomes (haploid). During the two successive meiotic divisions involved in the production of eggs, a primordial diploid egg cell is converted into a haploid egg and three small haploid polar bodies (minute cells). In this instance the egg receives far more cytoplasm than the polar bodies.

Multiple Fission

Some algae, some protozoans, and the true slime molds (Myxomycetes) regularly divide by multiple fission. In such cases the nucleus undergoes several mitotic divisions, producing a number of nuclei. After the nuclear divisions are complete, the cytoplasm separates, and each nucleus becomes encased in its own membrane to form an individual cell. In the Myxomycetes, the fusion of two haploid gametes or the fusion of two or more diploid zygotes (the structures that result from the union of two sex cells) results in the formation of a plasmodium—a motile, multinucleate mass of cytoplasm. The nuclei are in a syncytium, that is, there are no cell boundaries, and the nuclei flow freely in the motile plasmodium. As it feeds, the plasmodium enlarges, and the nuclei divide synchronously about once every 24 hours. The plasmodium may become very large, with millions of nuclei, but, ultimately, when conditions are right, it forms a series of small bumps, each of which becomes a small, fruiting body (a structure that bears the spores). During this process the nuclei undergo meiosis, and the final haploid nuclei are then isolated into uninucleate spores (reproductive bodies).

Many algae (e.g., the Siphonales and related groups) are multinucleate. In most instances the nuclei are in one common cytoplasm within a large and elaborate organism surrounded by a hard cell wall. As the wall becomes extended, the nuclei, which wander freely in the central cavity, undergo repeated mitoses. Again, either during the formation of zoospores (asexual reproductive cells) or after meiosis during gamete formation, a massive progressive division occurs. The most unusual of such organisms is the marine alga Acetabularia; many nuclei stay clumped together in one compound nucleus in the rootlike base, which often is as much as two inches (five centimetres) away from the tip of the plant. The compound nucleus breaks up just before gamete formation, and the minute individual nuclei undergo meiosis and wander to the elaborate tip structures, where they are released as uninucleate gametes.

Syncytial organisms raise the question of whether or not cells, in the strict sense, are necessary for the development of large organisms. Syncytia are also found in animals—e.g., in the early stages of development of fishes and insects—and in the voluntary muscles of man. The proposal of the

19th-century botanist Julius von Sachs is generally considered a satisfactory answer to this question; he suggested that the important matter was the existence not of a cell membrane but of a certain amount of cytoplasm surrounding a nucleus and acting as a unit of metabolism, which he called an energid. Cell reproduction, therefore, might be considered a special case of energid reproduction.

Reproduction of Organisms

In single-celled organisms (e.g., bacteria, protozoans, many algae, and some fungi), organismic and cell reproduction are synonymous, for the cell is the whole organism. Details of the process differ greatly from one form to the next and, if the higher ciliate protozoans are included, can be extraordinarily complex. It is possible for reproduction to be asexual, by simple division, or sexual. In sexual unicellular organisms the gametes can be produced by division (often multiple fission, as in numerous algae) or, as in yeasts, by the organism turning itself into a gamete and fusing its nucleus with that of a neighbour of the opposite sex, a process that is called conjugation. In ciliate protozoans (e.g., Paramecium), the conjugation process involves the exchange of haploid nuclei; each partner acquires a new nuclear apparatus, half of which is genetically derived from its mate. The parent cells separate and subsequently reproduce by binary fission. Sexuality is present even in primitive bacteria, in which parts of the chromosome of one cell can be transferred to another during mating.

Multicellular organisms also reproduce asexually and sexually; asexual, or vegetative, reproduction can take a great variety of forms. Many multicellular lower plants give off asexual spores, either aerial or motile and aquatic (zoospores), which may be uninucleate or multinucleate. In some cases the reproductive body is multicellular, as in the soredia of lichens and the gemmae of liverworts. Frequently, whole fragments of the vegetative part of the organism can bud off and begin a new individual, a phenomenon that is found in most plant groups. In many cases a spreading rhizoid (rootlike filament) or, in higher plants, a rhizome (underground stem) gives off new sprouts. Sometimes other parts of the plant have the capacity to form new individuals; for instance, buds of potentially new plants may form in the leaves; even some shoots that bend over and touch the ground can give rise to new plants at the point of contact.

Among animals, many invertebrates are equally well endowed with means of asexual reproduction. Numerous species of sponges produce gemmules, masses of cells enclosed in resistant cases, that can become new sponges. There are many examples of budding among coelenterates, the best known of which occurs in freshwater Hydra. In some species of flatworms, the individual worm can duplicate by pinching in two, each half then regenerating the missing half; this is a large task for the posterior portion, which lacks most of the major organs—brain, eyes, and pharynx. The highest animals that exhibit vegetative reproduction are the colonial tunicates (e.g., sea squirts), which, much like plants, send out runners in the form of stolons, small parts of which form buds that develop into new individuals. Vertebrates have lost the ability to reproduce vegetatively; their only form of organismic reproduction is sexual.

In the sexual reproduction of all organisms except bacteria, there is one common feature: haploid, uninucleate gametes are produced that join in fertilization to form a diploid, uninucleate zygote. At some later stage in the life history of the organism, the chromosome number is again reduced by meiosis to form the next generation of gametes. The gametes may be equal in size (isogamy), or one may be slightly larger than the other (anisogamy); the majority of forms have a large egg and a minute sperm (oogamy). The sperm are usually motile and the egg passive, except in higher plants, in which the sperm nuclei are carried in pollen grains that attach to the stigma (a female structure) of the flower and send out germ tubes that grow down to the egg nucleus in the ovary. Some organisms, such as most flowering plants, earthworms, and tunicates, are bisexual (hermaphroditic, or monoecious)—i.e., both the male and female gametes are produced by the same individual. All other organisms, including some plants (e.g., holly and the ginkgo tree) and all vertebrates, are unisexual (dioecious): the male and female gametes are produced by separate individuals.

Some sexual organisms partially revert to the asexual mode by a periodic degeneration of the sexual process. For instance, in aphids and in many higher plants the egg nucleus can develop into a new individual without fertilization, a kind of asexual reproduction that is called parthenogenesis.

Life-cycle Reproduction

Although organisms are often thought of only as adults, and reproduction is considered to be the formation of a new adult resembling the adult of the previous generation, a living organism, in reality, is an organism for its entire life cycle, from fertilized egg to adult, not for just one short part of that cycle. Reproduction, in these terms, is not just a stage in the life history of an organism but the organism's entire history. It has been pointed out that only the DNA of a cell is capable of replicating itself, and even that replication process requires specific enzymes that were themselves formed from DNA. Thus, the reproduction of all living forms must be considered in relation to time; what is reproduced is a series of copies that, like the sequence of individual frames of a motion picture, change through time in an exact and orderly fashion.

A few examples serve to illustrate the great variety of life cycles in living organisms. They also illustrate how different parts of the life cycle can change, and the fact that these changes are not confined solely to adult structures. One variation is that of minimum size—that is to say, the differences in the sizes of gametes (mature sex cells) and asexual bodies. An even greater variation in life cycles, however, involves maximum size; there is an enormous difference between a single-celled organism that divides by binary fission and a giant sequoia. Size is correlated with time. A bacterium requires about 30 minutes to complete its life history and divide in two (generation time); a giant sequoia bears its first cones and fertile seeds after 60 years. Not only is the life cycle of the sequoia 10,000,000 times longer than that of the bacterium, but the large difference in size also means that the tree must be elaborate and complex. It contains different tissue types that must be carefully duplicated from generation to generation.

Life Cycles of Plants

Most life histories, except perhaps for the simplest and smallest organisms, consist of different epochs. A large tree has a period of seed formation that involves many cell divisions after fertilization and the laying down of a small embryo in a hard resistant shell, or seed coat. There then follows a period of dormancy, sometimes prolonged, after which the seed germinates, and the adult form slowly emerges as the shoots and roots grow at the tips and the stem thickens. In some trees the leaves of the juvenile plant have a shape that is quite different from that of the taller, more mature individuals. Thus, even the growth phase may be subdivided into epochs, the final one being the flowering or gametebearing period. Some of the parasitic fungi have much more complex life histories. The wheat rust parasite, for example, has alternate hosts. While living on wheat, it produces two kinds of spores; it produces a third kind of spore when it invades its other host, the barberry, on which it winters and undergoes the sexual part of its life cycle.

In plants, variations in the epochs of the life cycle are often centred around the times of fertilization and meiosis. After fertilization the organism has the diploid number of chromosomes (diplophase); after meiosis it is haploid (haplophase). The two events vary in time with respect to each other. In some simple algae (e.g., Chlamydomonas), for example, most of the cycle is haploid; meiosis occurs immediately after fertilization. Yet in other algae, such as the sea lettuce (Ulva), two equal haploid and diploid cycles alternate. The outward morphological structures of mature Ulva are indistinguishable; the two cycles can be differentiated only by the size of the cell or nucleus, those of the haploid stage being half the size of those of the diploid stage.

In many of the higher algae, there is a progressive diminution of the haplophase and an increase in the importance of the diplophase, a trend that is especially noticeable in the evolution of the vascular plants (e.g., ferns, conifers, and flowering plants). In mosses, the haplophase, or gametophyte, is the main part of the green plant; the diplophase, or sporophyte, usually is a sporebearing spike that grows from the top of the plant. In ferns, the haplophase is reduced to a small, inconspicuous structure (prothallus) that grows in the damp soil; the large spore-bearing fern itself is entirely diploid. Finally, in higher plants the haploid tissue is confined to the ovary of the large diploid organism, a condition that is also prevalent in most animals.

Life Cycles of Animals

Invertebrate animals have a rich variety of life cycles, especially among those forms that undergo metamorphosis, a radical physical change. Butterflies, for instance, have a caterpillar stage (larva), a dormant chrysalis stage (pupa), and an adult stage (imago). One remarkable aspect of this development is that, during the transition from caterpillar to adult, most of the caterpillar tissue disintegrates and is used as food, thereby providing energy for the next stage of development, which begins when certain small structures (imaginal disks) in the larva start growing into the adult form. Thus, the butterfly undergoes essentially two periods of growth and development (larva and pupa–adult) and two periods of small size (fertilized egg and imaginal disks). A somewhat similar phenomenon is found in sea urchins; the larva, which is called a pluteus, has a small, wartlike bud that grows into the adult while the pluteus tissue disintegrates. In both examples it is as if the organism has two life histories, one built on the ruins of another.

Another life-cycle pattern found among certain invertebrates illustrates the principle that major differences between organisms are not always found in the physical appearance of the adult but in differences of the whole life history. In the coelenterate Obelia, for example, the egg develops into a colonial hydroid consisting of a series of branching Hydra-like organisms called polyps. Certain of these polyps become specialized (reproductive polyps) and bud off from the colony as free-swimming jellyfish (medusae) that bear eggs and sperm. As with caterpillars and sea urchins, two distinct phases occur in the life cycle of Obelia: the sessile (anchored), branched polyps and the motile medusae. In some related coelenterates the medusa form has been totally lost, leaving only the polyp stage to bear eggs and sperm directly. In still other coelenterates the polyp stage has been lost, and the medusae produce other medusae directly, without the sessile stage. There are, furthermore, intermediate forms between the extremes.

Natural Selection and Reproduction

The significance of biological reproduction can be explained entirely by natural selection. In formulating his theory of natural selection, Charles Darwin realized that, in order for evolution to occur, not only must living organisms be able to reproduce themselves but the copies must not all be identical; that is, they must show some variation. In this way the more successful variants would make a greater contribution to subsequent generations in the number of offspring. For such selection to act continuously in successive generations, Darwin also recognized that the variations had to be inherited, although he failed to fathom the mechanism of heredity. Moreover, the amount of variation is particularly important. According to what has been called the principle of compromise, which itself has been shaped by natural selection, there must not be too little or too much variation: too little produces no change; too much scrambles the benefit of any particular combination of inherited traits.

Of the numerous mechanisms for controlling variation, all of which involve a combination of checks and balances that work together, the most successful is that found in the large majority of all plants and animals—i.e., sexual reproduction. During the evolution of reproduction and variation, which are the two basic properties of organisms that not only are required for natural selection but are also subject to it, sexual reproduction has become ideally adapted to produce the right amount of variation and to allow new combinations of traits to be rapidly incorporated into an individual.

The Evolution of Reproduction

An examination of the way in which organisms have changed since their initial unicellular condition in primeval times shows an increase in multicellularity and therefore an increase in the size of both plants and animals. After cell reproduction evolved into multicellular growth, the multicellular organism evolved a means of reproducing itself that is best described as life-cycle reproduction. Size increase has been accompanied by many mechanical requirements that have necessitated a selection for increased efficiency; the result has been a great increase in the complexity of organisms. In terms of reproduction this means a great increase in the permutations of cell reproduction during the process of evolutionary development.

Size increase also means a longer life cycle, and with it a great diversity of patterns at different stages of the cycle. This is because each part of the life cycle is adaptive in that, through natural selection, certain characteristics have evolved for each stage that enable the organism to survive. The most extreme examples are those forms with two or more separate phases of their life cycle separated by a metamorphosis, as in caterpillars and butterflies; these phases may be shortened or extended by natural selection, as has occurred in different species of coelenterates.

To reproduce efficiently in order to contribute effectively to subsequent generations is another factor that has evolved through natural selection. For instance, an organism can produce vast quantities of eggs of which, possibly by neglect, only a small percent will survive. On the other hand, an organism can produce very few or perhaps one egg, which, as it develops, will be cared for, thereby greatly increasing its chances for survival. These are two strategies of reproduction; each has its advantages and disadvantages. Many other considerations of the natural history and structure of the organism determine, through natural selection, the strategy that is best for a particular species;

one of these is that any species must not produce too few offspring (for it will become extinct) or too many (for it may also become extinct by overpopulation and disease). The numbers of some organisms fluctuate cyclically but always remain between upper and lower limits. The question of how, through natural selection, numbers of individuals are controlled is a matter of great interest; clearly, it involves factors that influence the rate of reproduction.

The Evolution of Variation Control

Because inherited variation is largely handled by genes in the chromosomes, organisms that reproduce sexually require a single-cell stage in their life cycle, during which the haploid gamete of each parent can combine to form the diploid zygote. This is also often true in organisms that reproduce asexually, but in this case the asexual reproductive bodies (e.g., spores) are small and hence are effectively dispersed.

The amount of variation is controlled in a large number of ways, all of which involve a carefully balanced set of factors. These factors include whether the organism reproduces asexually or sexually; the mutation (gene change) rate; the number of chromosomes; the amount of exchange of parts of chromosomes (crossing over); the size of the individual (which correlates with complexity and generation time); the size of the population; the degree of inbreeding versus outbreeding; and the relative amounts and position of haploidy and diploidy in the life cycle. It is clear, therefore, that the mode of reproduction influences the amount of variation and vice versa; the two together permit natural selection to operate, and selection in turn modifies the mechanisms of reproduction and variation.

Asexual Reproduction

Asexual reproduction occurs when an organism makes more of itself without exchanging genetic information with another organism through sex.

In sexually reproducing organisms, the genomes of two parents are combined to create offspring with unique genetic profiles. This is beneficial to the population because genetically diverse populations have a higher chance of withstanding survival challenges such as disease and environmental changes.

Asexually reproducing organisms can suffer a dangerous lack of diversity – but they can also reproduce faster than sexually reproducing organisms, and a single individual can found a new population without the need for a mate.

Some organisms that practice asexual reproduction can exchange genetic information to promote diversity using forms of horizontal gene transfer such as bacteria who use plasmids to pass around small bits of DNA. However this method results in fewer unique genotypes than sexual reproduction.

Some species of plants, animals, and fungi are capable of both sexual and asexual reproduction, depending on the demands of the environment.

Asexual reproduction is practiced by most single-celled organisms including bacteria, archaebacteria, and protists. It is also practiced by some plants, animals, and fungi.

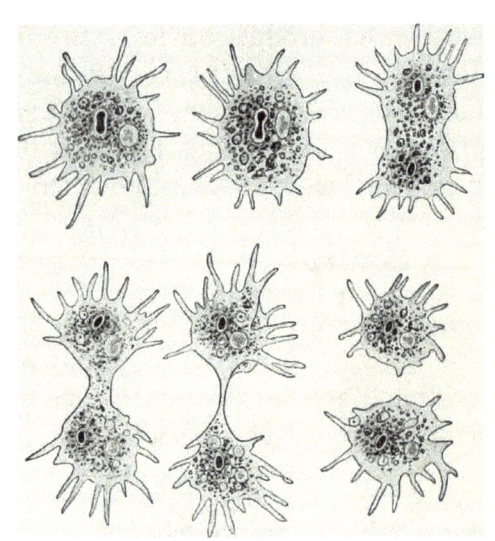

Evolution and animal life.

Advantages of Asexual Reproduction

1. Rapid population growth. This is especially useful for species whose survival strategy is to reproduce very fast.

Many species of bacteria, for example, can completely rebuild a population from just a single mutant individual in a matter of days if most members are wiped out by a virus.

2. No mate is needed to found a new population.

This is useful for species whose members may find themselves isolated, such as fungi that grow from wind-blown spores, plants that rely on pollinators for sexual reproduction, and animals inhabiting environments with low population density.

3. Lower resource investment. Asexual reproduction, which can often be accomplished just by having part of the parent organism split off and take on a life of its own, takes fewer resources than nurturing a new baby organism.

Many plants and sea creatures, for example, can simply cut a part of themselves off from the parent organism and have that part survive on its own.

Only offspring that are genetically identical to the parent can be produced in this way: nurturing the creation of a new organism whose tissue is different from the parents' tissue takes more time, energy, and resources.

This ability to simply split in two is one reason why asexual reproduction is faster than sexual reproduction.

Disadvantages of Asexual Reproduction

The biggest disadvantage of asexual reproduction is lack of diversity. Because members of an asexually reproducing population are genetically identical except for rare mutants, they are all susceptible to the same diseases, nutrition deficits, and other types of environmental hardships.

The Irish Potato Famine was one example of the down side of asexual reproduction: Ireland's potatoes, which had mainly reproduced through asexual reproduction, were all vulnerable when a potato-killing plague swept the island. As a result, almost all crops failed, and many people starved.

The near-extinction of the Gros-Michel banana is another example – one of two major cultivars of bananas, it became impossible to grow commercially in the 20th century after the emergence of a disease to which it was genetically vulnerable.

On the other hand, many species of bacteria actually take advantage of their high mutation rate to create some genetic diversity while using asexual reproduction to grow their colonies very rapidly. Bacteria have a higher rate of errors in copying genetic sequences, which sometimes leads to the creation of useful new traits even in the absence of sexual reproduction.

Types of Asexual Reproduction

There are many different ways to reproduce asexually. These include:

1. Binary fission. This method, in which a cell simply copies its DNA and then splits in two, giving a copy of its DNA to each "daughter cell," is used by bacteria and archaebacteria.

2. Budding. Some organisms split off a small part of themselves to grow into a new organism. This is practiced by many plants and sea creatures, and some single-celled eukaryotes such as yeast.

3. Vegetative propagation. Much like budding, this process involves a plant growing a new shoot which is capable of becoming a whole new organism. Strawberries are an example of plants that reproduce using "runners," which grow outward from a parent plant and later become separate, independent plants.

4. Sporogenesis. Sporogenesis is the production of reproductive cells, called spores, which can grow into a new organism.

Spores often use similar strategies to those of seeds. But unlike seeds, spores can be created without fertilization by a sexual partner. Spores are also more likely to spread autonomously, such as via wind, than to rely on other organisms such as animal carriers to spread.

5. Fragmentation. In fragmentation, a "parent" organism is split into multiple parts, each of which grows to become a complete, independent "offspring" organism. This process resembles budding and vegetative propagation, but with some differences.

For one, fragmentation may not be voluntary on the part of the "parent" organism. Earthworms and many plants and sea creatures are capable of regenerating whole organisms from fragments following injuries that split them into multiple pieces.

When fragmentation does occur voluntarily, the same parent organism may split into many roughly equal parts in order to form many offspring. This is different from the processes of budding and vegetative propagation, where an organism grows new parts which are small compared to the parent and which are intended to become offspring organisms.

6. Agamenogenesis. Agamenogenesis is the reproduction of normally sexual organisms without the need for fertilization. There are several ways in which this can happen.

In parthenogenesis, an unfertilized egg begins to develop into a new organism, which by necessity possesses only genes from its mother.

This occurs in a few species of all-female animals, and in females of some animal species when there are no males present to fertilize eggs.

In apomoxis, a normally sexually reproducing plant reproduces asexually, producing offspring that are identical to the parent plant, due to lack of availability of a male plant to fertilize female gametes.

In nucellar embryony, an embryo is formed from a parents' own tissue without meiosis or the use of reproductive cells. This is primarily known to occur in citrus fruit, which may produce seeds in this way in the absence of male fertilization.

Examples of Asexual Reproduction

Bacteria

All bacteria reproduce through asexual reproduction, by splitting into two "daughter" cells that are genetically identical to their parents.

Some bacteria can undergo horizontal gene transfer – in which genetic material is passed "horizontally" from one organism to another, instead of "vertically" from parent to child. Because they have only one cell, bacteria are able to change their genetic material as mature organisms.

The process of genetic exchange between bacterial cells is sometimes referred to as "sex," although it is performed to change the genotype of a mature bacterium, not as a means of reproduction.

Bacteria can afford to use this survival strategy because their extremely rapid reproduction makes harmful genetic mutations – such as copying errors or horizontal gene transfer gone wrong – inconsequential to the whole population. As long as a few individuals survive mutation and calamity, those individuals will be able to rebuild the bacterial population quickly.

This strategy of "reproduce fast, mutate often" is a major reason why bacteria are so quick to develop antibiotic resistance. They have also been seen to "invent" whole new biochemistries in the lab, such as one species of bacteria that spontaneously acquired the ability to perform anaerobic respiration.

This strategy would not work well for an organism that invests highly in the survival of individuals, such as multicellular organisms.

Slime Molds

Slime molds are a fascinating organism that sometimes behave like a multicellular organism, and sometimes behave like a colony of single-celled organisms.

Unlike animals, plants, and fungi, the cells in a slime mold are not bound together in a fixed shape and dependent on each other for survival. The cells that make up a slime mold are capable of living individually and may spread or separate when food is abundant, much like individuals in a colony of bacteria.

But slime mold cells are eukaryotic, and can display a high degree of cooperation to the point of creating a temporary extracellular matrix and a "body" which may become large and complex. Slime molds whose cells are working cooperatively can be mistaken for fungi, and can perform locomotion.

Slime molds can produce spores much like a fungus, and they can also reproduce through fragmentation. Environmental causes or injury may cause a slime mold to disperse into many parts, and units as small as a single cell may grow into a whole new slime mold colony/organism.

New Mexico Whiptail Lizards

This species of lizard was created by the hybridization of two neighboring species. Genetic incompatibility between the hybrid parents made it impossible for healthy males to be born: however, the female hybrids were capable of parthenogenesis, making them a reproductively independent population.

All New Mexico whiptail lizards are female. New members of the species can be created through hybridization of the parent species, or through parthenogenesis by female New Mexico whiptails.

Possibly as a remnant of their sexually reproducing past, New Mexico whiptail lizards do have a "mating" behavior which they must go through to reproduce. Members of this species are "mated with" by other members, and the lizard playing the female role will go onto lay eggs.

It is thought that the mating behavior stimulates ovulation, which can then result in a parthenogenic pregnancy. The lizard playing the "male" role in the courtship does not lay eggs.

Sexual Reproduction

Common frogs in amplexus. Sexually reproducing individuals spend a considerable amount of time and energy locating mates, exchanging genetic material, and often caring for young.

Sexual reproduction is a method for producing a new individual organism while combining genes from two parents. A single sperm and egg fuse during fertilization , and their genomes combine in the new zygote . Sperm are small and contain little more than the father's genes. Eggs are large and contain the mother's genes and all cellular components necessary for the early development and nutrition of the embryo. The sexual dimorphism in gamete size is echoed in many other traits of adults and has resulted in the evolution of different male and female reproductive strategies. Sexual reproduction is widespread in almost all groups of multicellular organisms.

The Stages of Sexual Reproduction

The various stages of sexual reproduction are:

1) Pre-fertilization

This stage involves the events prior to fertilization. Gamete formation (gametogenesis) and transfer of gamete are the two processes that take place during this stage. Gametes are sex cells which are haploid (23 chromosomes) in nature and are distinct in males and females.

Male gamete is called sperm whereas female gamete is called ovum or egg. In every organism, these gametes form within special structures. Since female gamete is immobile, male gametes need to be transferred for fertilization. In plants, this is pre-fertilization happens through pollination. Unisexual animals transfer gametes by sexual intercourse.

2) Fertilization

The process in which the haploid male and female gametes meet and fuse together to form a zygote is fertilization or syngamy. This can occur either outside the body known as External fertilization or inside the body known as Internal fertilization.

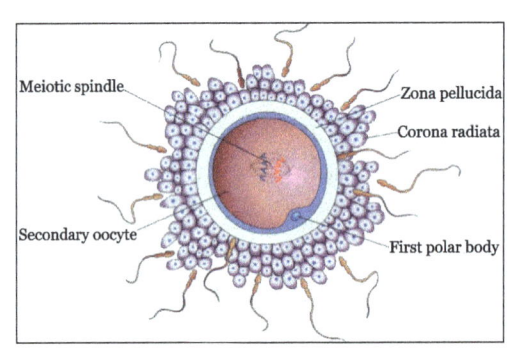

3) Post-fertilization

Fertilization results in diploid zygote formation. Eventually, the zygote divides mitotically and develops into an embryo. This process is called embryogenesis. During embryogenesis, cells differentiate and modify accordingly. Zygote development depends on the organism and its life cycle.

Animals are classified into oviparous and viviparous based on whether the zygote develops outside or inside the body respectively. In angiosperms, zygote develops into the ovary and ovary transforms into fruit while ovules develop into seeds.

Life Cycles of Sexually Reproducing Organisms

In sexual reproduction, the genetic material of two individuals is combined to produce genetically diverse offspring that differ from their parents. Fertilization and meiosis alternate in sexual life cycles. What happens between these two events depends upon the organism. The process of meiosis, the division of the contents of the nucleus that divides the chromosomes among gametes, reduces the chromosome number by half, while fertilization, the joining of two haploid gametes, restores the diploid condition. There are three main categories of life cycles in eukaryotic organisms: diploid-dominant, haploid-dominant, and alternation of generations.

Diploid-Dominant Life Cycle

In the diploid-dominant life cycle, the multicellular diploid stage is the most obvious life stage, as occurs with most animals, including humans. Nearly all animals employ a diploid-dominant life cycle strategy in which the only haploid cells produced by the organism are the gametes. Early in the development of the embryo, specialized diploid cells, called germ cells, are produced within the gonads (e.g. testes and ovaries). Germ cells are capable of mitosis to perpetuate the cell line and meiosis to produce gametes. Once the haploid gametes are formed, they lose the ability to divide again. There is no multicellular haploid life stage. Fertilization occurs with the fusion of two gametes, usually from different individuals, restoring the diploid state.

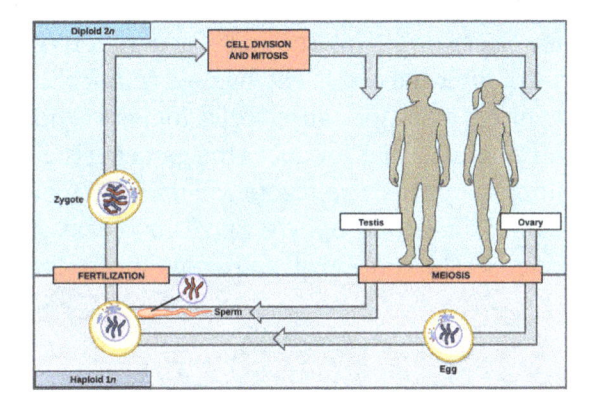

Diploid-Dominant Life Cycle: In animals, sexually-reproducing adults form haploid gametes from diploid germ cells. Fusion of the gametes gives rise to a fertilized egg cell, or zygote. The zygote will undergo multiple rounds of mitosis to produce a multicellular offspring. The germ cells are generated early in the development of the zygote.

Haploid-Dominant Life Cycle

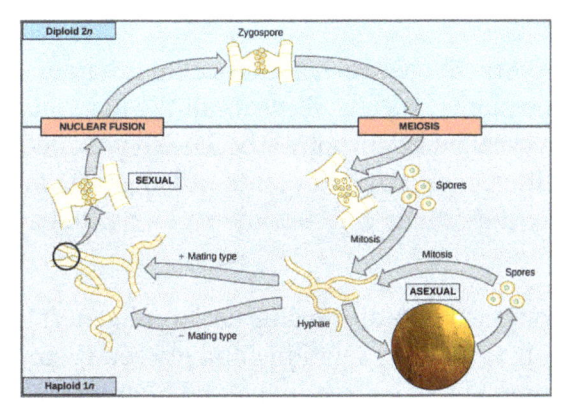

Haploid-Dominant Life Cycle: Fungi, such as black bread mold (Rhizopus nigricans), have haploid-dominant life cycles. The haploid multicellular stage produces specialized haploid cells by mitosis that fuse to form a diploid zygote. The zygote undergoes meiosis to produce haploid spores. Each spore gives rise to a multicellular haploid organism by mitosis.

Within haploid-dominant life cycles, the multicellular haploid stage is the most obvious life stage. Most fungi and algae employ a life cycle type in which the "body" of the organism, the ecologically important part of the life cycle, is haploid. The haploid cells that make up the tissues of the dominant multicellular stage are formed by mitosis. During sexual reproduction, specialized haploid cells from two individuals, designated the (+) and (−) mating types, join to form a diploid zygote. The zygote immediately undergoes meiosis to form four haploid cells called spores. Although haploid like the "parents," these spores contain a new genetic combination from two parents. The spores can remain dormant for various time periods. Eventually, when conditions are conducive, the spores form multicellular haploid structures by many rounds of mitosis.

Alternation of Generations

The third life-cycle type, employed by some algae and all plants, is a blend of the haploid-dominant and diploid-dominant extremes. Species with alternation of generations have both haploid and diploid multicellular organisms as part of their life cycle. The haploid multicellular plants are called gametophytes because they produce gametes from specialized cells. Meiosis is not directly involved in the production of gametes because the organism that produces the gametes is already a haploid. Fertilization between the gametes forms a diploid zygote. The zygote will undergo many rounds of mitosis and give rise to a diploid multicellular plant called a sporophyte. Specialized cells of the sporophyte will undergo meiosis and produce haploid spores. The spores will subsequently develop into the gametophytes.

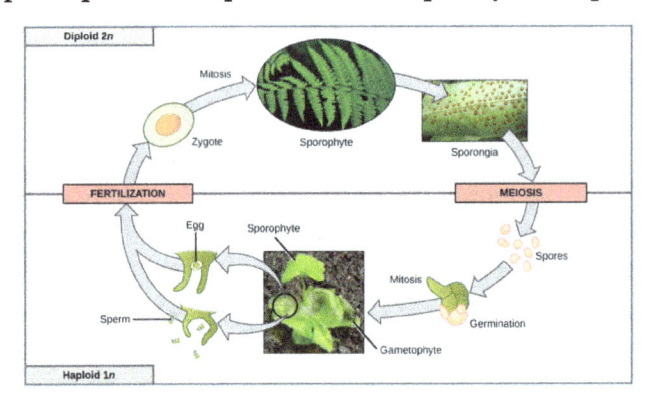

Plants have a life cycle that alternates between a multicellular haploid organism and a multicellular diploid organism. In some plants, such as ferns, both the haploid and diploid plant stages are free-living. The diploid plant is called a sporophyte because it produces haploid spores by meiosis. The spores develop into multicellular, haploid plants called gametophytes because they produce gametes. The gametes of two individuals will fuse to form a diploid zygote that becomes the sporophyte.

Although all plants utilize some version of the alternation of generations, the relative size of the sporophyte and the gametophyte and the relationship between them vary greatly. In plants such as moss, the gametophyte organism is the free-living plant, while the sporophyte is physically dependent on the gametophyte. In other plants, such as ferns, both the gametophyte and sporophyte plants are free-living; however, the sporophyte is much larger. In seed plants, such as magnolia trees and daisies, the gametophyte is composed of only a few cells and, in the case of the female gametophyte, is completely retained within the sporophyte.

Sexual reproduction takes many forms in multicellular organisms. However, at some point in each type of life cycle, meiosis produces haploid cells that will fuse with the haploid cell of another organism. The mechanisms of variation (crossover, random assortment of homologous chromosomes, and random fertilization) are present in all versions of sexual reproduction. The fact that nearly every multicellular organism on earth employs sexual reproduction is strong evidence for the benefits of producing offspring with unique gene combinations, although there are other possible benefits as well.

Reproduction in Plants

Plant reproduction is the process by which plants generate new individuals, or offspring. Reproduction is either sexual or asexual. Sexual reproduction is the formation of offspring by the fusion of gametes. Asexual reproduction is the formation of offspring without the fusion of gametes. Sexual reproduction results in offspring genetically different from the parents. Asexual offspring are genetically identical except for mutation. In higher plants, offspring are packaged in a protective seed, which can be long lived and can disperse the offspring some distance from the parents. In flowering plants (angiosperms), the seed itself is contained inside a fruit, which may protect the developing seeds and aid in their dispersal.

Asexual Reproduction in Plants

The mode of reproduction by which new plants are produced without the reproductive unit of plants – flowers. This mode of reproduction occurs without the fusion of male and female gametes. Asexual reproduction produces new plants, which are the copies of the mother plant.

Asexual reproduction occurs in different kinds which includes the budding, fragmentation, vegetative propagation, and spore formation.

Types of Asexual Reproduction

Budding

Budding is the mode of asexual reproduction, wherein a new plant is developed from an outgrowth plant, called a bud. A bud is generally formed due to cell division at one particular site.

For example, if you keep a potato for a long time, you can notice a number of small growths, which are commonly referred to as 'eyes'. Each of them can be planted which will grow up like a clone of an original potato plant.

Vegetative Propagation

It is any form of asexual reproduction occurring in plants, in which new plants are produced from the vegetative parts of the plants, i.e. roots, stems or buds. Vegetative propagation in plants can occur both by naturally or also can be artificially induced by horticulturists.

The most common techniques of vegetative propagation are:

- Stems – Runners are the stems which usually grow in a horizontal form above the ground. They have the nodes where the buds are formed. These buds usually grow into a new plant.
- Roots – A new plant is developed from around, inflamed, modified roots called tubers. Example: Sweet Potato.
- Leaves – In some plants, detached leaves from the parent plant can be used to grow a new plant. They exhibit growth of small plants, called plantlets, on the edge of their leaves. Example: Bryophyllum.

Fragmentation

This is a mode of asexual reproduction in which a new plant is produced from a portion of the parent plant. Each section or a part of the plants develop into a mature, fully grown individual. Some plants possess specialized structures for reproduction through fragmentation. This type of reproduction happens naturally where the small part of the plant fall off onto soil and then begin to grow up into a new plant. This mode is often used by nurseries and greenhouses to produce plants quickly.

Spore Formation

Many plants and algae form spores in their life cycle. A spore is an asexual reproductive body, surrounded by a hard protective cover to withstand unfavorable conditions such as high temperature and low humidity. Under favorable conditions, the spores germinate and grow into new plants. Plants like moss and ferns use this mode of reproduction.

Sexual Reproduction in Flowering Plants

The type of reproduction in which the fusion of male and female gamete occurs is known as sexual reproduction. Flowers produce male and female gametes in the reproductive organs.

Reproductive Part of a Plant

Flowers are the reproductive parts of a plant. Male and female gametes are produced in the flower.

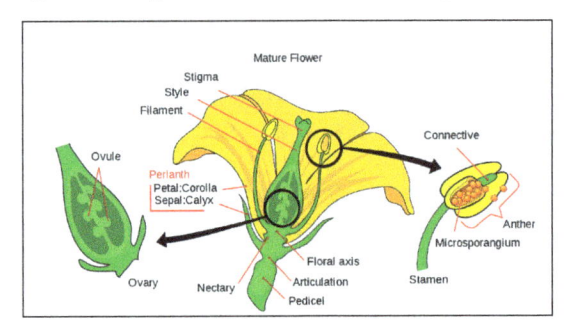

The flower consists of four main parts-

- Sepals: They are the green leaf-like structures which protect the flower in the bud stage.
- Petals: They are the coloured structures. They are the most beautiful part of the flower that

attracts the birds and insects for pollination.

- Stamens: They are the male reproductive parts. Stamens consist of two parts – anther and filament. Anther is a swollen part of the stamen. The filament is a long, slender stalk which attaches the stamen to the flower. Pollen grains are formed inside the anther which produces the male gametes.
- Pistil or Carpel: It is the female reproductive part of the flower. It consists of three parts – stigma, style and the ovary. The stigma is at the top of the pistil. It receives the pollen grains during pollination. The style is a long, tube-like structure which passes the pollen grains to the ovary. The ovary is the swollen part present at the basal part of the pistil. The ovary consists of the ovule which bears the female gamete.

Uni-sexual and Bisexual Flowers

Flowers can be of two types depending upon whether both the sexual parts – male and female are present in the same flower or different flower.

- Bisexual flowers: Flowers which contain both the stamens and the pistil are known as bisexual flowers. For example- Rose, Hibiscus, Gulmohar, Mustard etc.
- Uni-sexual flowers: The flowers which contain either the stamens or the pistil are known as unisexual flowers. For example- Maize, Papaya, Cucumber etc.

Pollination

Transfer of pollen grains from the anthers to the receptive stigma is known as pollination. Pollination is of two types:

- Self-pollination: In this process pollen grains from the anthers pollinate the stigma of the same flower.
- Cross-pollination: In this case pollen grains of one flower fall on a different flower of either the same plant or another plant.

Agents of Pollination

Pollination occurs through some external agents like:

1) Wind

The pollen grains which are very light in weight can be passed on by wind from one flower to another or to the stigma of the same flower. This is known as wind pollination. Examples of wind pollination are pollination in wheat, rice, maize etc. Wind-pollinated flowers have the following characteristics:

- Wind-pollinated flowers are mostly small and not showy.
- They produce lightweight pollens in large number and do not produce nectar.
- The stigma is sticky.

2) Insects

When an insect sits on the flowers to collect the nectar, the pollen grains stick on their legs and wings. When they travel to another flower, they carry the pollens with them. In this way, they carry out pollination. Examples are a sweet pea, orchids, sunflower, buttercup. Characteristics of insect-pollinated flowers:

- These flowers have bright colours, are large and showy.
- They have nectaries, sticky stigma and sticky pollen grains.
- Some birds like Sunbird and hummingbirds and mammals like Squirrel also help in pollination.

3) Water

Aquatic plants carry out water pollination. In this, the pollen grains are released in water passively and are carried out by water currents to the other flowers for pollination. Examples of water-pollination are seagrass, hydrilla etc.

Fertilization

Fertilization is the fusion of male and female gametes. It results in the formation of single-celled zygote.

Germination of Pollen Grain

The pollen grain develops pollen tube through the style and reaches the ovary. It enters the ovule from a small opening. Pollen tube carries the male gamete. Male gamete fuses with the female gamete to form the zygote.

Post-fertilization Change

Following changes take place in the flower after fertilization. The fertilized zygote grows into an embryo. The embryo has two parts – plumule which grows into shoot and radical which grows into the root.

- The ovary grows into fruit.
- The other parts of the flower fall off.
- The ovule develops into a seed.

Double Fertilization

Double fertilization is a complex fertilization mechanism of angiosperms which involves the fusion of a female gametophyte with two male gametes. One male gamete fuses with the female egg whereas the second sperm fuses with two polar nuclei to give rise to a triploid body that forms an endosperm.

Double Fertilization in Angiosperms

Angiosperms are flower-bearing plants surrounding us and these flowers are their reproductive units which include male and female reproductive organs. Each contains gametes – sperm and egg cells respectively.

Pollination helps the pollen grains to reach stigma via style. Two sperm cells enter the ovule-synergid cell. This proceeds to fertilization.

In angiosperms, fertilization results in two structures namely zygote and endosperm; thus fertilization is known as double fertilization.

By definition, double fertilization is a complex fertilization method where out of two sperm cells, one fuses with the egg cell and other fuses with two polar nuclei which result in diploid (2n) zygote and triploid (3n) primary endosperm nucleus (PEN) respectively.

Since endosperm is a product of the fusion of three haploid nuclei, it is called triple fusion. Eventually, the primary endosperm nucleus develops into the primary endosperm cell (PEC) and then into the endosperm.

The zygote becomes an embryo after numerous cell divisions.

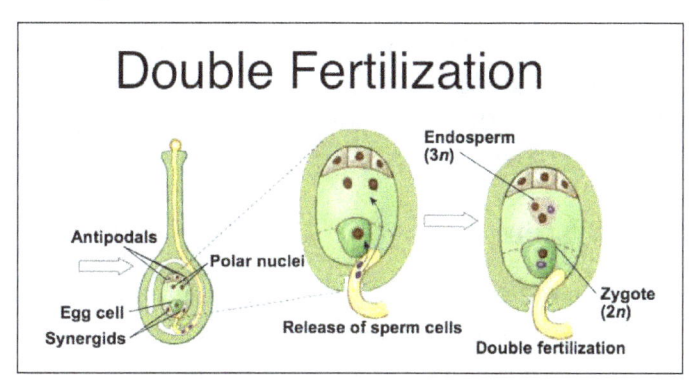

Development of Embryo

Once fertilization is done embryonic development starts and no more sperm can enter the ovary. The fertilized ovule develops into a seed and ovary tissues develop as fleshy fruit which encloses the seed.

The first stage of embryonic development includes division of the zygote into the upper terminal cell and lower basal cell.

The basal cell develops into suspensor which helps in nutrition transport to the growing embryo. The terminal cell develops into proembryo. Further embryonic development into seed takes in an interval of time.

Significance of Double Fertilization

The significance of double fertilization is mentioned below:

- Produces a Food Supply for the Seed: An embryo is produced within the seed during the process of double fertilization. This procedure is also used as an additional food source by some plants.

- Provides Extra Protection to the Species: Double fertilization increases the life span of a flowering plant. There are two male gametes. Therefore the chances of them fusing with the female gametes are very high. The endosperm formed after fertilization stores nutrients. So, the plant does not have to allocate energy-intensive resources for the unfertilized egg.

- Energy Conservation: When the embryo starts coming into shape, it starts conserving energy due to double fertilization. The essential supply is found in the seed itself.

Allows Rapid Seed Development

Since two male gametes are involved in double fertilization, the development of seed is more rapid as compared to other forms of reproduction. The nucleus of the endosperm divides quickly to structure the nutrient-rich tissue, which inturn enhances the development of embryo.

Reproduction in Humans

Reproduction in human beings is by sexual reproduction where both the male and female gametes fertilize to give rise to an embryo. The fertilization of human embryo occurs inside the body of the female. Thus, it is called Internal Fertilization. Human Beings are viviparous organisms who give rise to Embryos directly instead of laying eggs.

Human Reproductive System

Human reproductive system is the organ system by which humans reproduce and bear live offspring. Provided all organs are present, normally constructed, and functioning properly, the essential features of human reproduction are (1) liberation of an ovum, or egg, at a specific time in the reproductive cycle, (2) internal fertilization of the ovum by spermatozoa, or sperm cells, (3) transport of the fertilized ovum to the uterus, or womb, (4) implantation of the blastocyst, the early embryo developed from the fertilized ovum, in the wall of the uterus, (5) formation of a placenta and maintenance of the unborn child during the entire period of gestation, (6) birth of the child and expulsion of the placenta, and (7) suckling and care of the child, with an eventual return of the maternal organs to virtually their original state.

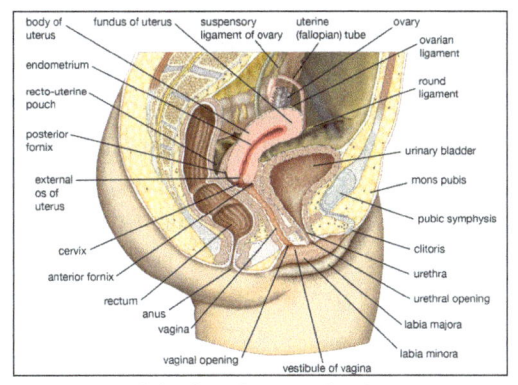

Organs of the female reproductive system.

For this biological process to be carried out, certain organs and structures are required in both the male and the female. The source of the ova (the female germ cells) is the female ovary; that of spermatozoa (the male germ cells) is the testis. In females, the two ovaries are situated in the pelvic cavity; in males, the two testes are enveloped in a sac of skin, the scrotum, lying below and outside the abdomen. Besides producing the germ cells, or gametes, the ovaries and testes are the source of hormones that cause full development of secondary sexual characteristics and also the proper functioning of the reproductive tracts. These tracts comprise the fallopian tubes, the uterus, the vagina, and associated structures in females and the penis, the sperm channels (epididymis, ductus deferens, and ejaculatory ducts), and other related structures and glands in males. The function of the fallopian tube is to convey an ovum, which is fertilized in the tube, to the uterus, where gestation (development before birth) takes place. The function of the male ducts is to convey spermatozoa from the testis, to store them, and, when ejaculation occurs, to eject them with secretions from the male glands through the penis.

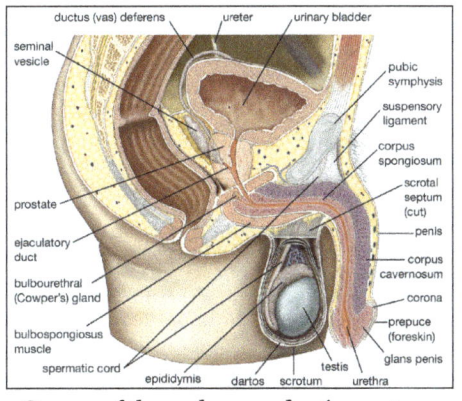

Organs of the male reproductive system.

At copulation, or sexual intercourse, the erect penis is inserted into the vagina, and spermatozoa contained in the seminal fluid (semen) are ejaculated into the female genital tract. Spermatozoa then pass from the vagina through the uterus to the fallopian tube to fertilize the ovum in the outer part of the tube. Females exhibit a periodicity in the activity of their ovaries and uterus, which starts at puberty and ends at the menopause. The periodicity is manifested by menstruation at intervals of about 28 days; important changes occur in the ovaries and uterus during each reproductive, or menstrual, cycle. Periodicity, and subsequently menstruation, is suppressed during pregnancy and lactation.

Development of the Reproductive Organs

The sex of a child is determined at the time of fertilization of the ovum by the spermatozoon. The differences between a male and a female are genetically determined by the chromosomes that each possesses in the nuclei of the cells. Once the genetic sex has been determined, there normally follows a succession of changes that will result, finally, in the development of an adult male or female. There is, however, no external indication of the sex of an embryo during the first eight weeks of its life within the uterus. This is a neutral or indifferent stage during which the sex of an embryo can be ascertained only by examination of the chromosomes in its cells.

The next phase, one of differentiation, begins first in gonads that are to become testes and a week or so later in those destined to be ovaries. Embryos of the two sexes are initially alike in possessing similar duct systems linking the undifferentiated gonads with the exterior and in having similar external genitalia, represented by three simple protuberances. The embryos each have four ducts, the subsequent fate of which is of great significance in the eventual anatomical differences between men and women. Two ducts closely related to the developing urinary system are called mesonephric, or wolffian, ducts. In males each mesonephric duct becomes differentiated into four related structures: a duct of the epididymis, a ductus deferens, an ejaculatory duct, and a seminal vesicle. In females the mesonephric ducts are largely suppressed. The other two ducts, called the paramesonephric or müllerian ducts, persist, in females, to develop into the fallopian tubes, the uterus, and part of the vagina; in males they are largely suppressed. Differentiation also occurs in the primitive external genitalia, which in males become the penis and scrotum and in females the vulva (the clitoris, labia, and vestibule of the vagina).

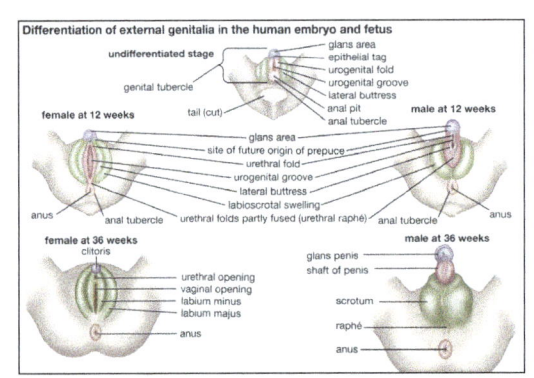

Differentiation of the external genitalia in the human embryo and fetus.

At birth the organs appropriate to each sex have developed and are in their adult positions but are not functioning. Various abnormalities can occur during development of sex organs in embryos, leading to hermaphroditism, pseudohermaphroditism, and other chromosomally induced conditions. During childhood until puberty there is steady growth in all reproductive organs and a gradual development of activity. Puberty marks the onset of increased activity in the sex glands and the steady development of secondary sexual characteristics.

In males at puberty the testes enlarge and become active, the external genitalia enlarge, and the capacity to ejaculate develops. Marked changes in height and weight occur as hormonal secretion from the testes increases. The larynx, or voice box, enlarges, with resultant deepening of the voice. Certain features in the skeleton, as seen in the pelvic bones and skull, become accentuated. The hair in the armpits and the pubic hair becomes abundant and thicker. Facial hair develops, as well as hair on the chest, abdomen, and limbs. Hair at the temples recedes. Skin glands become more active, especially apocrine glands (a type of sweat gland that is found in the armpits and groin and around the anus).

In females at puberty, the external genitalia enlarge and the uterus commences its periodic activity with menstruation. The breasts develop, and there is a deposition of body fat in accordance with the usual contours of the mature female. Growth of axillary (armpit) and pubic hair is more abundant, and the hair becomes thicker.

The Male Reproductive System

The male gonads are the testes; they are the source of spermatozoa and also of male sex hormones called androgens. The other genital organs are the epididymides; the ductus, or vasa, deferentia; the seminal vesicles; the ejaculatory ducts; and the penis; as well as certain accessory structures, such as the prostate and the bulbourethral (Cowper) glands. The principal functions of these structures are to transport the spermatozoa from the testes to the exterior, to allow their maturation on the way, and to provide certain secretions that help form the semen.

External Genitalia

The Penis

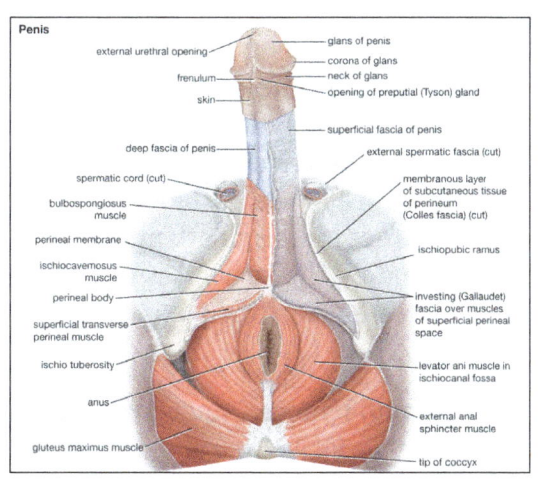

The human penis.

The penis, the male organ of copulation, is partly inside and partly outside the body. The inner part, attached to the bony margins of the pubic arch (that part of the pelvis directly in front and at the base of the trunk), is called the root of the penis. The second, or outer, portion is free, pendulous, and enveloped all over in skin; it is termed the body of the penis. The organ is composed chiefly of cavernous or erectile tissue that becomes engorged with blood to produce considerable enlargement and erection. The penis is traversed by a tube, the urethra, which serves as a passage both for urine and for semen.

The body of the penis, sometimes referred to as the shaft, is cylindrical in shape when flaccid but when erect is somewhat triangular in cross section, with the angles rounded. This condition arises because the right corpus cavernosum and the left corpus cavernosum, the masses of erectile tissue, lie close together in the dorsal part of the penis, while a single body, the corpus spongiosum, which contains the urethra, lies in a midline groove on the undersurface of the corpora cavernosa. The dorsal surface of the penis is that which faces upward and backward during erection.

The slender corpus spongiosum reaches beyond the extremities of the erectile corpora cavernosa and at its outer end is enlarged considerably to form a soft, conical, sensitive structure called the glans penis. The base of the glans has a projecting margin, the corona, and the groove where the corona overhangs the corpora cavernosa is referred to as the neck of the penis. The glans is traversed by the urethra, which ends in a vertical, slitlike, external opening. The skin over the penis is thin and loosely adherent and at the neck is folded forward over the glans for a variable distance to form the prepuce or foreskin. A median fold, the frenulum of the prepuce, passes to the undersurface of the glans to reach a point just behind the urethral opening. The prepuce can usually be readily drawn back to expose the glans.

The root of the penis comprises two crura, or projections, and the bulb of the penis. The crura and the bulb are attached respectively to the edges of the pubic arch and to the perineal membrane (the fibrous membrane that forms a floor of the trunk). Each crus is an elongated structure covered by the ischiocavernosus muscle, and each extends forward, converging toward the other, to become continuous with one of the corpora cavernosa. The oval bulb of the penis lies between the two crura and is covered by the bulbospongiosus muscle. It is continuous with the corpus spongiosum. The urethra enters it on the flattened deep aspect that lies against the perineal membrane, traverses its substances, and continues into the corpus spongiosum.

The two corpora cavernosa are close to one another, separated only by a partition in the fibrous sheath that encloses them. The erectile tissue of the corpora is divided by numerous small fibrous bands into many cavernous spaces, relatively empty when the penis is flaccid but engorged with blood during erection. The structure of the tissue of the corpus spongiosum is similar to that of the corpora cavernosa, but there is more smooth muscle and elastic tissue. A deep fascia, or sheet of connective tissue, surrounding the structures in the body of the penis is prolonged to form the suspensory ligament, which anchors the penis to the pelvic bones at the midpoint of the pubic arch.

The penis has a rich blood supply from the internal pudendal artery, a branch of the internal iliac artery, which supplies blood to the pelvic structures and organs, the buttocks, and the inside of the thighs. Erection is brought about by distension of the cavernous spaces with blood, which is prevented from draining away by compression of the veins in the area.

The penis is amply supplied with sensory and autonomic (involuntary) nerves. Of the autonomic nerve fibres the sympathetic fibres cause constriction of blood vessels, and the parasympathetic fibres cause their dilation. It is usually stated that ejaculation is brought about by the sympathetic system, which at the same time inhibits the desire to urinate and also prevents the semen from entering the bladder.

The Scrotum

The scrotum is a pouch of skin lying below the pubic symphysis and just in front of the upper parts of the thighs. It contains the testes and lowest parts of the spermatic cord. A scrotal septum or partition divides the pouch into two compartments and arises from a ridge, or raphe, visible on the outside of the scrotum. The raphe turns forward onto the undersurface of the penis and is continued back onto the perineum (the area between the legs and as far back as the anus). This arrangement indicates the bilateral origin of the scrotum from two genital swellings that lie one on each side of the base of the phallus, the precursor of the penis or clitoris in the embryo. The swellings are also referred to as the labioscrotal swellings, because in females they remain separate to form the labia majora and in males they unite to form the scrotum.

The skin of the scrotum is thin, pigmented, devoid of fatty tissue, and more or less folded and wrinkled. There are some scattered hairs and sebaceous glands on its surface. Below the skin is a layer of involuntary muscle, the dartos, which can alter the appearance of the scrotum. On exposure of the scrotum to cold air or cold water, the dartos contracts and gives the scrotum a shortened, corrugated appearance; warmth causes the scrotum to become smoother, flaccid, and less closely tucked in around the testes. Beneath the dartos muscle are layers of fascia continuous with those forming the coverings of each of the two spermatic cords, which suspend the testes within the scrotum and contain each ductus deferens, the testicular blood and lymph vessels, the artery to the cremaster muscle (which draws the testes upward), the artery to each ductus deferens, the genital branch of the genitofemoral nerve, and the testicular network of nerves.

The scrotum is supplied with blood by the external pudendal branches of the femoral artery, which is the chief artery of the thigh, and by the scrotal branches of the internal pudendal artery. The veins follow the arteries. The lymphatic drainage is to the lymph nodes in the groin.

The Testes

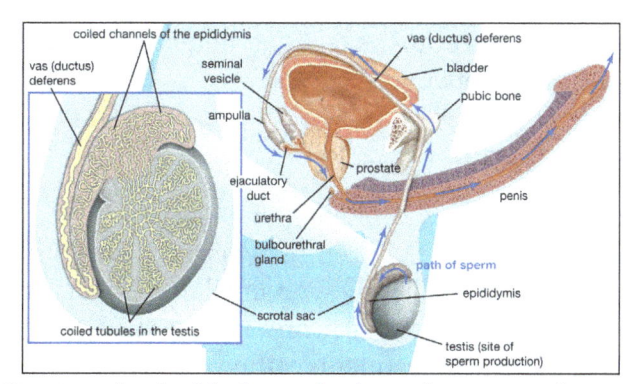

Structures involved in the production and transport of semen.

The two testes, or testicles, which usually complete their descent into the scrotum from their point of origin on the back wall of the abdomen in the seventh month after conception, are suspended in the scrotum by the spermatic cords. Each testis is 4 to 5 cm (about 1.5 to 2 inches) long and is enclosed in a fibrous sac, the tunica albuginea. This sac is lined internally by the tunica vasculosa, containing a network of blood vessels, and is covered by the tunica vaginalis, which is a continuation of the membrane that lines the abdomen and pelvis. The tunica albuginea has extensions into each testis that act as partial partitions to divide the testis into approximately 250 compartments, or lobules.

Each lobule contains one or more convoluted tubules, or narrow tubes, where sperm are formed. The tubules, if straightened, would extend about 70 cm (about 28 inches). The multistage process of sperm formation, which takes about 60 days, goes on in the lining of the tubules, starting with the spermatogonia, or primitive sperm cells, in the outermost layer of the lining. Spermatozoa (sperm) leaving the tubules are not capable of independent motion, but they undergo a further maturation process in the ducts of the male reproductive tract; the process may be continued when, after ejaculation, they pass through the female tract. Maturation of the sperm in the female tract is called capacitation.

Each spermatozoon is a slender elongated structure with a head, a neck, a middle piece, and a tail. The head contains the cell nucleus. When the spermatozoon is fully mature, it is propelled by the lashing movements of the tail.

The male sex hormone testosterone is produced by Leydig cells. These cells are located in the connective (interstitial) tissue that holds the tubules together within each lobule. The tissue becomes markedly active at puberty under the influence of the interstitial-cell-stimulating hormone of the anterior lobe of the pituitary gland; this hormone in women is called luteinizing hormone. Testosterone stimulates the male accessory sex glands (prostate, seminal vesicles) and also brings about the development of male secondary sex characteristics at puberty. The hormone may also be necessary to cause maturation of sperm and to heighten the sex drive of the male. The testis is also the source of some of the female sex hormone estrogen, which may exert an influence on pituitary activity.

Each testis is supplied with blood by the testicular arteries, which arise from the front of the aorta just below the origin of the renal (kidney) arteries. Each artery crosses the rear abdominal wall, enters the spermatic cord, passes through the inguinal canal, and enters the upper end of each testis at the back. The veins leaving the testis and epididymis form a network, which ascends into the spermatic cord. The lymph vessels, which also pass through the spermatic cord, drain to the lateral and preaortic lymph nodes. Nerve fibres to the testis accompany the vessels; they pass through the renal and aortic nerve plexuses, or networks.

Structures of the Sperm Canal

The epididymis, ductus deferens (or vas deferens), and ejaculatory ducts form the sperm canal. Together they extend from the testis to the urethra, where it lies within the prostate. Sperm are conveyed from the testis along some 20 ductules, or small ducts, which pierce the fibrous capsule to enter the head of the epididymis. The ductules are straight at first but become dilated and then much convoluted to form distinct compartments within the head of the epididymis. They each open into a single duct, the highly convoluted duct of the epididymis, which constitutes the "body" and "tail" of the structure. It is held together by connective tissue but if unraveled would be nearly 6 metres (20 feet) long. The duct enlarges and becomes thicker-walled at the lower end of the tail of the epididymis, where it becomes continuous with the ductus deferens.

The ductules from the testis have a thin muscular coat and a lining that consists of alternating groups of high columnar cells with cilia (hairlike projections) and low cells lacking cilia. The cilia assist in moving sperm toward the epididymis. In the duct of the epididymis the muscle coat is thicker and the lining is thick with tufts of large nonmotile cilia. There is some evidence that the ductules and the first portion of the duct of the epididymis remove excess fluid and extraneous debris from the testicular secretions entering these tubes. The blood supply to the epididymis is by a branch from the testicular artery given off before that vessel reaches the testis.

The ductus deferens, or vas deferens, is the continuation of the duct of the epididymis. It commences at the lower part of the tail of the epididymis and ascends along the back border of the testis to its upper pole. Then, as part of the spermatic cord, it extends to the deep inguinal ring. Separating from the other elements of the spermatic cord—the blood vessels, nerves, and lymph vessels—at the ring, the ductus deferens makes its way through the pelvis toward the base of the prostate, where it is joined by the seminal vesicle to form the ejaculatory duct. A part of the ductus that is dilated and rather tortuous, near the base of the urinary bladder, is called the ampulla.

The ductus deferens has a thick coat of smooth muscle that gives it a characteristic cordlike feel. The longitudinal muscle fibres are well developed, and peristaltic contractions (contractions in waves) move the sperm toward the ampulla. The mucous membrane lining the interior is in longitudinal folds and is mostly covered with nonciliated columnar cells, although some cells have nonmotile cilia. The ampulla is thinner-walled and probably acts as a sperm store.

Accessory Organs

The Prostate Gland, Seminal Vesicles and Bulbourethral Glands

These structures provide secretions to form the bulk of the seminal fluid of an ejaculate. The prostate gland is in the lesser or true pelvis, centred behind the lower part of the pubic arch. It lies in front of the rectum. The prostate is shaped roughly like an inverted pyramid; its base is directed upward and is immediately continuous with the neck of the urinary bladder. The urethra traverses its substance. The two ejaculatory ducts enter the prostate near the upper border of its posterior surface. The prostate is of a firm consistency, surrounded by a capsule of fibrous tissue and smooth muscle. It measures about 4 cm across, 3 cm in height, and 2 cm front to back (about 1.6 by 1.2 by 0.8 inch) and consists of glandular tissue contained in a muscular framework. It is imperfectly divided into three lobes. Two lobes at the side form the main mass and are continuous behind the urethra. In front of the urethra they are connected by an isthmus of fibromuscular tissue devoid of glands. The third, or median, lobe is smaller and variable in size and may lack glandular tissue. There are three clinically significant concentric zones of prostatic glandular tissue about the urethra. A group of short glands that are closest to the urethra and discharge mucus into its channel are subject to simple enlargement. Outside these is a ring of submucosal glands (glands from which the mucosal glands develop), and farther out is a large outer zone of long branched glands, composing the bulk of the glandular tissue. Prostate cancer is almost exclusively confined to the outer zone. The glands of the outer zone are lined by tall columnar cells that secrete prostatic fluid under the influence of androgens from the testis. The fluid is thin, milky, and slightly acidic.

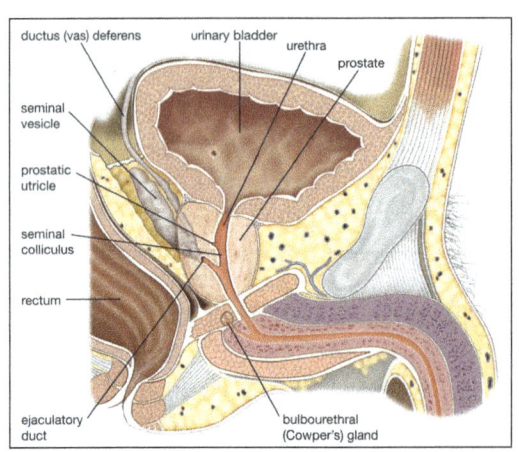

Sagittal section of the male reproductive organs, showing the prostate gland and seminal vesicles.

The seminal vesicles are two structures, about 5 cm (2 inches) in length, lying between the rectum and the base of the bladder. Their secretions form the bulk of semen. Essentially, each vesicle consists of a much-coiled tube with numerous diverticula or outpouches that extend from the main tube, the whole being held together by connective tissue. At its lower end the tube is constricted to form a straight duct or tube that joins with the corresponding ductus deferens to form the ejaculatory duct. The vesicles are close together in their lower parts, but they are separated above where they lie close to the deferent ducts. The seminal vesicles have longitudinal and circular layers of smooth muscle, and their cavities are lined with mucous membrane, which is the source of the secretions of the organs. These secretions are ejected by muscular contractions during ejaculation. The activity of the vesicles is dependent on the production of the hormone androgen by the testes. The secretion is thick, sticky, and yellowish; it contains the sugar fructose and is slightly alkaline.

The bulbourethral glands, often called Cowper glands, are pea-shaped glands that are located beneath the prostate gland at the beginning of the internal portion of the penis. The glands, which measure only about 1 cm (0.4 inch) in diameter, have slender ducts that run forward and toward the centre to open on the floor of the spongy portion of the urethra. They are composed of a network of small tubes, or tubules, and saclike structures; between the tubules are fibres of muscle and elastic tissue that give the glands muscular support. Cells within the tubules and sacs contain droplets of mucus, a thick protein compound. The fluid excreted by these glands is clear and thick and acts as a lubricant; it is also thought to function as a flushing agent that washes out the urethra before the semen is ejaculated; it may also help to make the semen less watery and to provide a suitable living environment for the sperm.

Ejaculatory Ducts

The two ejaculatory ducts lie on each side of the midline and are formed by the union of the duct of the seminal vesicle, which contributes secretions to the semen, with the end of the ductus deferens at the base of the prostate. Each duct is about 2 cm (about 0.8 inch) long and passes between a lateral and the median lobe of the prostate to reach the floor of the prostatic urethra. This part of the urethra has on its floor (or posterior wall) a longitudinal ridge called the urethral crest. On each side is a depression, the prostatic sinus, into which open the prostatic ducts. In the middle of the urethral crest is a small elevation, the colliculus seminalis, on which the opening of the prostatic utricle is found. The prostatic utricle is a short diverticulum or pouch lined by mucous membrane;

it may correspond to the vagina or uterus in the female. The small openings of the ejaculatory ducts lie on each side of or just within the opening of the prostatic utricle. The ejaculatory ducts are thin-walled and lined by columnar cells.

The Female Reproductive System

The female gonads, or sexual glands, are the ovaries; they are the source of ova (eggs) and of the female sex hormones estrogens and progestogens. The fallopian, or uterine, tubes conduct ova to the uterus, which lies within the lesser or true pelvis. The uterus connects through the cervical canal with the vagina. The vagina opens into the vestibule about which lie the external genitalia, collectively known as the vulva.

External Genitalia

The female external genitalia include the structures placed about the entrance to the vagina and external to the hymen, the membrane across the entrance to the vagina. They are the mons pubis (also called the mons veneris), the labia majora and minora, the clitoris, the vestibule of the vagina, the bulb of the vestibule, and the greater vestibular glands.

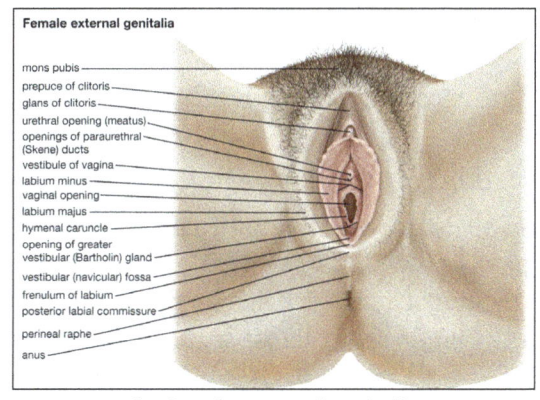

The female external genitalia.

The mons pubis is the rounded eminence, made by fatty tissue beneath the skin, lying in front of the pubic symphysis. A few fine hairs may be present in childhood; later, at puberty, they become coarser and more numerous. The upper limit of the hairy region is horizontal across the lower abdomen.

The labia majora are two marked folds of skin that extend from the mons pubis downward and backward to merge with the skin of the perineum. They form the lateral boundaries of the vulval or pudendal cleft, which receives the openings of the vagina and the urethra. The outer surface of each labium is pigmented and hairy; the inner surface is smooth but possesses sebaceous glands. The labia majora contain fat and loose connective tissue and sweat glands. They correspond to the scrotum in the male and contain tissue resembling the dartos muscle. The round ligament ends in the tissue of the labium. The labia minora are two small folds of skin, lacking fatty tissue, that extend backward on each side of the opening into the vagina. They lie inside the labia majora and are some 4 cm (about 1.5 inches) in length. In front, an upper portion of each labium minus passes over the clitoris—the structure in the female corresponding to the penis (excluding the urethra) in the male—to form a fold, the prepuce of the clitoris, and a lower portion passes beneath the clitoris to form its frenulum. The two labia minora are joined at the back across the midline by a fold that becomes stretched at childbirth. The labia minora lack hairs but possess sebaceous and sweat glands.

The clitoris is a small erectile structure composed of two corpora cavernosa separated by a partition. Partially concealed beneath the forward ends of the labia minora, it possesses a sensitive tip of spongy erectile tissue, the glans clitoridis. The external opening of the urethra is some 2.5 cm (about 1 inch) behind the clitoris and immediately in front of the vaginal opening.

The vestibule of the vagina is the cleft between the labia minora into which the urethra and vagina open. The hymen vaginae lies at the opening of the vagina: it is a thin fold of mucous membrane that varies in shape. After rupture of the hymen, the small rounded elevations that remain are known as the carunculae hymenales. The bulb of the vestibule, corresponding to the bulb of the penis, is two elongated masses of erectile tissue that lie one on each side of the vaginal opening. At their posterior ends lie the greater vestibular glands, small mucous glands that open by a duct in the groove between the hymen and each labium minus. They correspond to the bulbourethral glands of the male.

The blood supply and nerve supply of the female external genital organs are similar to those supplying corresponding structures in the male.

Internal Structures

The Vagina

The vagina (the word means "sheath") is the canal that extends from the cervix (outer end) of the uterus within the lesser pelvis down to the vestibule between the labia minora. The orifice of the vagina is guarded by the hymen. The vagina lies behind the bladder and urethra and in front of the rectum and anal canal. Its walls are collapsed; the anterior wall is some 7.5 cm (3 inches) in length, whereas the posterior wall is about 1.5 cm (0.6 inch) longer. The vagina is directed obliquely upward and backward. The axis of the vagina forms an angle of over 90° with that of the uterus. This angle varies considerably depending on conditions in the bladder, in the rectum, and during pregnancy. The cervix of the uterus projects for a short distance into the vagina and is normally pressed against its posterior wall. There are, therefore, recesses in the vagina at the back, on each side, and at the front of the cervix. These are known as the posterior fornix (behind the cervix and the largest), the lateral fornices (at the sides), and the anterior fornix (at the front of the cervix).

The upper part of the posterior wall of the vagina is covered by peritoneum or membrane that is folded back onto the rectum to form the recto-uterine pouch. The lower part of the posterior vaginal wall is separated from the anal canal by a mass of tissue known as the perineal body.

The vagina has a mucous membrane and an outer smooth muscle coat closely attached to it. The mucous membrane has a longitudinal ridge in the midline of both the anterior and posterior walls. The ridges are known as the columns of the vagina; many rugae, or folds, extend from them to each side. The furrows between the rugae are more marked on the posterior wall and become especially pronounced before the birth of a child. The membrane undergoes little change during the menstrual cycle (except in its content of glycogen, a complex starchlike carbohydrate); this is in contradistinction to the situation in many mammals in which marked exfoliation (shedding of the surface cells) can occur. No glands are present in the vaginal lining, and mucus present has been secreted by the glands in the cervical canal of the uterus. The smooth muscle coat consists of an outer longitudinal layer and a less developed inner circular layer. The lower part of the vagina is surrounded by the bulbospongiosus muscle, a striped muscle attached to the perineal body.

The blood supply to the vagina is derived from several adjacent vessels, there being a vaginal artery from the internal iliac artery and also vaginal branches from the uterine, middle rectal, and internal pudendal arteries, all branches of the internal iliac artery. The nerve supply to the lower part of the vagina is from the pudendal nerve and from the inferior hypogastric and uterovaginal plexuses.

The Uterus

Uterine Structure

The uterus, or womb, is shaped like an inverted pear. It is a hollow, muscular organ with thick walls, and it has a glandular lining called the endometrium. In an adult the uterus is 7.5 cm (3 inches) long, 5 cm (2 inches) in width, and 2.5 cm (1 inch) thick, but it enlarges to four to five times this size in pregnancy. The narrower, lower end is called the cervix; this projects into the vagina. The cervix is made of fibrous connective tissue and is of a firmer consistency than the body of the uterus. The two fallopian tubes enter the uterus at opposite sides, near its top. The part of the uterus above the entrances of the tubes is called the fundus; the part below is termed the body. The body narrows toward the cervix, and a slight external constriction marks the juncture between the body and the cervix.

The uterus does not lie in line with the vagina but is usually turned forward (anteverted) to form approximately a right angle with it. The position of the uterus is affected by the amount of distension in the urinary bladder and in the rectum. Enlargement of the uterus in pregnancy causes it to rise up into the abdominal cavity, so that there is closer alignment with the vagina. The nonpregnant uterus also curves gently forward; it is said to be anteflexed. The uterus is supported and held in position by the other pelvic organs, by the muscular floor or diaphragm of the pelvis, by certain fibrous ligaments, and by folds of peritoneum. Among the supporting ligaments are two double-layered broad ligaments, each of which contains a fallopian tube along its upper free border and a round ligament, corresponding to the gubernaculum testis of the male, between its layers. Two ligaments—the cardinal (Mackenrodt) ligaments—at each side of the cervix are also important in maintaining the position of the uterus.

The cavity of the uterus is remarkably small in comparison with the size of the organ. Except during pregnancy, the cavity is flattened, with front and rear walls touching, and is triangular. The triangle is inverted, with its base at the top, between the openings of the two fallopian tubes, and with its apex at the isthmus of the uterus, the opening into the cervix. The canal of the cervix is flattened from front to back and is somewhat larger in its middle part. It is traversed by two longitudinal ridges and has oblique folds stretching from each ridge in an arrangement like the branches of a tree. The cervical canal is 2.5 cm (about 1 inch) in length; its opening into the vagina is called the external os of the uterus. The external os is small, almost circular, and often depressed. After childbirth, the external os becomes bounded by lips in front and in back and is thus more slitlike.

The cervical canal is lined by a mucous membrane containing numerous glands that secrete a clear, alkaline mucus. The upper part of this lining undergoes cyclical changes resembling, but not as marked as, those occurring in the body of the uterus. Numerous small cysts (nabothian cysts) are found in the cervical mucous membrane. It is from this region that cervical smears are taken in order to detect early changes indicative of cancer.

The uterus is composed of three layers of tissue. On the outside is a serous coat of peritoneum (a membrane exuding a fluid like blood minus its cells and the clotting factor fibrinogen), which partially covers the organ. In front it covers only the body of the cervix; behind it covers the body and the part of the cervix that is above the vagina and is prolonged onto the posterior vaginal wall; from there it is folded back to the rectum. At the side the peritoneal layers stretch from the margin of the uterus to each side wall of the pelvis, forming the two broad ligaments of the uterus.

The middle layer of tissue (myometrium) is muscular and comprises the greater part of the bulk of the organ. It is very firm and consists of densely packed, unstriped, smooth muscle fibres. Blood vessels, lymph vessels, and nerves are also present. The muscle is more or less arranged in three layers of fibres running in different directions. The outermost fibres are arranged longitudinally. Those of the middle layer run in all directions without any orderly arrangement; this layer is the thickest. The innermost fibres are longitudinal and circular in their arrangement.

The innermost layer of tissue in the uterus is the mucous membrane, or endometrium. It lines the uterine cavity as far as the isthmus of the uterus, where it becomes continuous with the lining of the cervical canal. The endometrium contains numerous uterine glands that open into the uterine cavity and are embedded in the cellular framework or stroma of the endometrium. Numerous blood vessels and lymphatic spaces are also present. The appearances of the endometrium vary considerably at the different stages in reproductive life. It begins to reach full development at puberty and thereafter exhibits dramatic changes during each menstrual cycle. It undergoes further changes before, during, and after pregnancy, during the menopause, and in old age. These changes are for the most part hormonally induced and controlled by the activity of the ovaries.

The Endometrium in the Menstrual Cycle

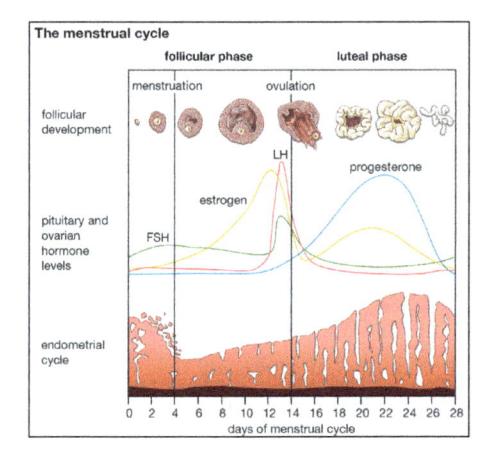

To understand the nature of the changes in the endometrium during each menstrual cycle it is usual to consider the endometrium to be composed of three layers. They blend imperceptibly but are functionally distinct: the inner two layers are shed at menstruation, and the outer or basal layer remains in position against the innermost layer of the myometrium. The three layers are called, respectively, the stratum compactum, the stratum spongiosum, and the stratum basale epidermidis. The stratum compactum is nearest to the uterine cavity and contains the lining cells and the necks of the uterine glands; its stroma is relatively dense. Superficial blood vessels lie beneath the lining cells. The stratum spongiosum is the large middle layer. It contains the main portions of uterine glands and accompanying blood vessels; the stromal cells are more loosely arranged and larger than in the stratum compactum. The stratum basale epidermidis lies against the uterine muscle; it contains blood vessels and the bases of the uterine glands. Its stroma remains relatively unaltered during the menstrual cycle.

The menstrual cycle extends over a period of about 28 days (normal range 21–34 days), from the first day of one menstrual flow to the first day of the next. It reflects the cycle of changes occurring in the ovary, which is itself under the control of the anterior lobe of the pituitary gland. The menstrual cycle is divided into four phases: menstrual, postmenstrual, proliferative, and secretory.

The secretory phase reaches its climax about a week after ovulation. Ovulation occurs in midcycle, about 14 days before the onset of the next menstrual flow. The endometrium has been prepared and has been stimulated to a state of active secretion for the reception of a fertilized ovum. The stage has been set for the attachment of the blastocyst, derived from a fertilized ovum, to the endometrium and for its subsequent embedding. This process is called implantation; its success depends on the satisfactory preparation of the endometrium in both the proliferative and secretory phases. When implantation occurs, a hormone from certain cells of the blastocyst causes prolongation of the corpus luteum and its continued activity. This causes suppression of menstruation and results in the maintenance of the endometrium and its further stimulation by progesterone, with consequent increased thickening. The endometrium of early pregnancy is known as the decidua.

In a cycle in which fertilization of the ovum has not taken place, the secretory phase terminates in menstruation.

The endometrium needs to be in a certain state of preparedness before implantation can occur. When this stage has been passed, menstruation occurs. Repair then reestablishes an endometrium capable of being stimulated again to the critical stage when implantation can occur.

Blood Supply and Innervation

The uterus is supplied with blood by the two uterine arteries, which are branches of the internal iliac arteries, and by ovarian arteries, which connect with the ends of the uterine arteries and send branches to supply the uterus. The nerves to the uterus include the sympathetic nerve fibres, which produce contraction of uterine muscle and constriction of vessels, and parasympathetic (sacral) fibres, which inhibit muscle activity and cause dilation of blood vessels.

The Fallopian Tubes

The fallopian, or uterine, tubes carry ova from the ovaries to the cavity of the uterus. Each opens into the abdominal cavity near an ovary at one end and into the uterus at the other. Three sections of the tubes are distinguished: the funnel-shaped outer end, or infundibulum; the expanded and thin-walled intermediate portion, or ampulla; and the cordlike portion, the isthmus, that opens into the uterus. The infundibulum is fringed with irregular projections called fimbriae. One fimbria, somewhat larger than the others, is usually attached to the ovary. The opening into the abdomen is at the bottom of the infundibulum and is small. Fertilization of the ovum usually occurs in the ampulla of the tube. Normally the fertilized egg is transported to the uterus, but occasionally it may adhere to the tube and start developing as an ectopic pregnancy, or tubal pregnancy. The tube is unable to support this pregnancy, and the conceptus may be extruded through the abdominal opening or may cause rupture of the tube, with ensuing hemorrhage.

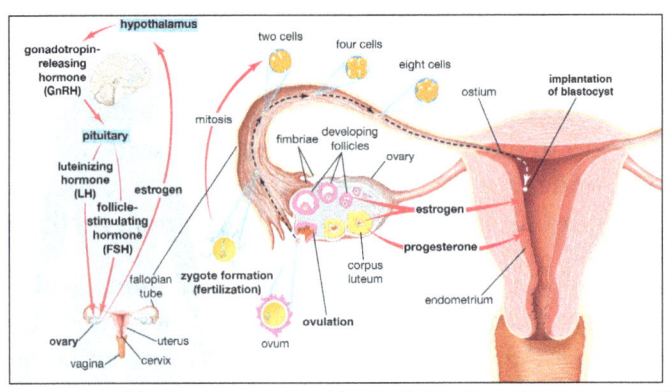

Major structures and hormones involved in the initiation of pregnancy. Also seen, at right, is the development of an egg cell (ovum) from follicle to embryo.

The fallopian tube is covered by peritoneum except on its border next to the broad ligament. There are inner circular and outer longitudinal layers of smooth muscle fibres continuous with those of the uterus. The inner lining has numerous longitudinal folds that are covered with ciliated columnar and secretory cells. Muscular contraction, movement of the hairlike cilia, and the passage of the watery secretions all probably assist in the transport of sperm to the ampulla and of a fertilized ovum toward the uterus.

The Ovaries

Ovarian Structure

The female gonads, or primary sex organs, corresponding to the testes in a male, are the two ovaries. Each is suspended by a mesentery, or fold of membrane, from the back layer of the broad ligament of the uterus. In a woman who has not been pregnant, the almond-shaped ovary lies in a vertical position against a depression, the ovarian fossa, on the side wall of the lesser pelvis. This relationship is altered during and after pregnancy. Each ovary is somewhat over 2.5 cm (1 inch) in length, 1.25 cm (0.5 inch) across, and slightly less in thickness, but the size varies much with age and with state of activity.

The mesentery of the ovary helps to keep it in position, and within this membrane lie the ovarian artery and vein, lymphatic vessels, and nerve fibres. The fallopian tube arches over the ovary and curves downward on its inner or medial surface.

Except at its hilum, the point where blood vessels and the nerve enter the ovary and where the mesentery is attached, the surface of the ovary is smooth and is covered by cubical cells. Beneath the surface, the substance of the ovary is divided into an outer portion, the cortex, and an inner portion, or medulla. The outermost part of the cortex, immediately beneath the outer covering, forms a thin connective tissue zone, the tunica albuginea. The rest of the cortex consists of stromal or framework cells, contained in a fine network of fibres, and also the follicles and corpora lutea.

The ovarian follicles, sometimes called graafian follicles, are rounded enclosures for the developing ova in the cortex near the surface of the ovary. At birth and in childhood they are present as numerous primary or undeveloped ovarian follicles. Each contains a primitive ovum, or oocyte, and each is covered by a single layer of flattened cells. As many as 700,000 primary follicles are contained in the two ovaries of a young female. Most of these degenerate before or after puberty.

Ovulation

During the onset of puberty and thereafter until menopause (except during pregnancy), there is a cyclic development of one or more follicles each month into a mature follicle. The covering layer of the primary follicle thickens and can be differentiated into an inner membrana granulosa and an outer vascularized theca interna. The cells of these layers (mostly the theca interna) produce estrogenic steroid hormones that exert their effects on the endometrium of the uterus and on other tissues. The maintenance and growth of the follicle to maturity is brought about by a follicle-stimulating hormone (FSH) from the anterior lobe of the pituitary gland. Another hormone, called luteinizing hormone (LH), from the anterior lobe, assists FSH to cause the maturing, now fluid-filled follicle to secrete estrogens. LH also causes a ripe follicle (1.0–1.5 cm [0.4–0.6 inch] in diameter) to rupture, causing the liberation of the oocyte into the peritoneal cavity and thence into the fallopian tube. This liberation of the oocyte is called ovulation; it occurs at about the midpoint of the reproductive cycle, on the 13th or 14th day of a 28-day cycle as measured from the first day of the menstrual flow.

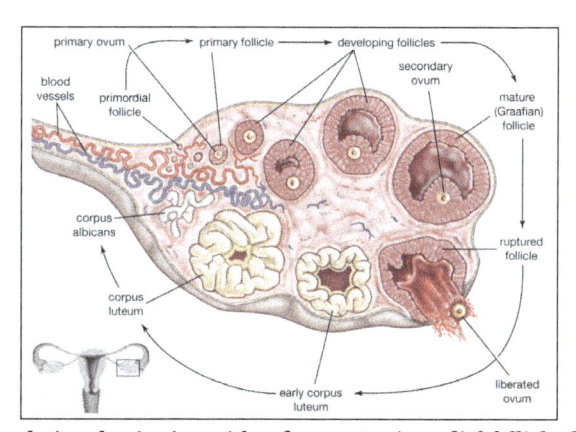

The steps of ovulation, beginning with a dormant primordial follicle that grows and matures and is eventually released from the ovary into the fallopian tube.

After ovulation the ruptured follicle collapses because of loss of its follicular fluid and rapidly becomes transformed into a soft, well-vascularized glandular structure known as the corpus luteum("yellow body"). The corpus luteum develops rapidly, becomes vascularized after about four days, and is fully established by nine days. The gland produces the steroid hormone progesterone and some estrogens. Its activity is both stimulated and maintained by luteinizing hormone. Progesterone stimulates glandular proliferation and secretion in an endometrium primed by estrogens.

While the ovarian follicle matures, the primary oocyte divides into a secondary oocyte and a small rudimentary ovum called the first polar body. This occurs at about the time when the follicle develops its cavity; the oocyte also gains a translucent acellular covering, or envelope, the zona pellucida. The secondary oocyte is liberated at ovulation; it is 120–140 micrometres in diameter and is surrounded by the zona pellucida and a few layers of cells known as the corona radiata. The final maturation of the oocyte, with the formation of the rudimentary ovum called the second polar body, occurs at the time of fertilization.

If fertilization does not occur, then the life of the corpus luteum is limited to about 14 days. Degeneration of the gland starts toward the end of this period, and menstruation occurs. The corpus luteum shrinks, fibrous tissue is formed, and it is converted into a scarlike structure called a corpus albicans, which persists for a few months.

Should fertilization occur and be followed by implantation of the blastocyst, hormones (particularly human chorionic gonadotropin) are produced by cells of the blastocyst to prolong the life of the corpus luteum. It persists in an active state for at least the first two months of pregnancy, until the placental tissue has taken over its hormone-producing function. The corpus luteum of pregnancy then also retrogresses, becoming a fibrous scar by the time of parturition.

Blood Supply and Innervation

The ovarian arteries arise from the front of the aorta in a manner similar to the testicular arteries, but at the brim of the lesser pelvis they turn down into the pelvic cavity. Passing in the suspensory ligament of the ovary, each artery reaches the broad ligament below the fallopian tube and then passes into the mesovarium to divide into branches distributed to the ovary. One branch continues in the broad ligament to anastomose with the uterine artery. The ovarian veins emerge from each ovary as a network that eventually becomes a single vein; the terminations are similar to those of the testicular veins. The nerves are derived from the ovarian nerve network on the ovarian artery.

Gametogenesis

Gametogenesis is the process whereby a haploid cell (n) is formed from a diploid cell (2n) through meiosis and cell differentiation. Gametogenesis in the male is known as spermatogenesis and produces spermatozoa. Gametogenesis in the female is known as oogenesis and result in the formation of ova.

Spermatogenesis

Males start producing sperm when they reach puberty, which is usually from 10-16 years old. Sperm are produced in large quantities (~200 million a day) to maximise the likelihood of sperm reaching the egg. Sperm are continually produced as males need to be ready to utilise the small window of fertility of the female.

Sperm production occurs in the testes of the male, specifically in the seminiferous tubules. The tubules are kept separate from the systemic circulation by the blood-testis barrier.

The blood-testis barrier is formed by Sertoli cells and is important in preventing hormones and constituents of the systemic circulation from affecting the developing sperm, and also in preventing the immune system of the male from recognising the sperm as foreign – as the sperm are genetically different from the male and will express different surface antigens. Sertoli cells also have a role in supporting the developing spermatozoa.

Spermatogonia are the initial pool of diploid cell that divide by mitosis to give two identical cells. One of these cells will be used to replenish the pool of spermatogonia – these cells are A1 spermatogonia. This replenishment of spermatogonia means that males are fertile throughout their adult life. The other cell – type B spermatogonium – will eventually form mature sperm.

Type B spermatogonia replicate by mitosis several times to form identical diploid cells linked by cytoplasm bridges, these cells are now known as primary spermatocytes. Primary spermatocytes then undergo meiosis.

- Meiosis I produces two haploid cells known as secondary spermatocytes.
- Meiosis II produces four haploid cells known as Spermatids.

The cytoplasmic bridges break down and the spermatids are released into the lumen of the seminiferous tubule – a process called spermiation. The spermatids undergo spermiogenesis (remodelling and differentiation into mature spermatozoa) as they travel along the seminiferous tubules until they reach the epididymis.

From the seminiferous tubule they travel to the rete testis, which acts to "concentrate" the sperm by removing excess fluid, before moving to the epididymis where the sperm is stored and undergoes the final stages of maturation.

Spermatogenesis takes approximately 70 days, therefore in order for sperm production to be continuous and not intermittent, multiple spermatogenic processes are occurring simultaneously within the same seminiferous tubule, with new groups of spermatogonia arising every 16 days (spermatogenic cycle). Each of these populations of spermatogenic cells will be at different stages of spermatogenesis.

Note that once sperm leave the male body and enter the female reproductive tract, the conditions there cause the sperm to undergo capacitation, which is the removal of cholesterol and glycoproteins from the head of the sperm cell to allow it to bind to the zona pellucida of the egg cell.

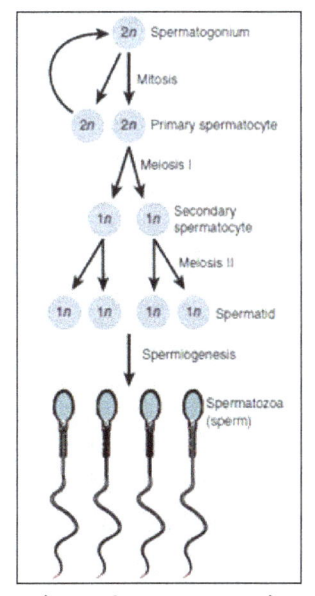

Figure: Spermatogenesis.

Oogenesis

Oogenesis differs from spermatogenesis in that it begins in the foetus prior to birth. Primordial germ cells (which originate in the yolk sac of the embryo) move to colonise the cortex of the primordial gonad and replicate by mitosis to peak at approximately 7 million by mid-gestation (~20 weeks). Cell death occurs after this peak to leave 2 million cells which begin meiosis I before birth and are known as primary oocytes. Therefore, a human female is born with approximately 2 million primary oocytes arrested in meiosis and these make up a finite supply of potential ova.

The primary oocytes are arranged in the gonads in clusters surrounded by flattened epithelial cells called follicular cells and these form primordial follicles. The primary oocytes are arrested in prophase stage of meiosis I.

During childhood, further atresia (cell death) occurs, leaving ~40,000 eggs at puberty.

Once puberty begins, a number of primary oocytes (15-20) begin to mature each month, although only one of these reaches full maturation to become an oocyte.

The primary oocytes undergo 3 stages:

- Pre-antral,
- Antral,
- Preovulatory.

Pre-antral Stage

The primary oocyte grows dramatically whilst still being arrested in meiosis I. The follicular cells grow and proliferate to form a stratified cuboidal epithelium. These cells are now known as granulosa cells and secrete glycoproteins to form the zona pellucida around the primary oocyte. Surrounding connective tissue cells also differentiates to become the theca folliculi, a specialised layer of surrounding cells that is responsive to LH and can secrete androgens under its influence.

Antral Stage

Fluid filled spaces form between granulosa cells, these eventually combine together to form a central fluid filled space called the antrum. The follicles are now called secondary follicles. In each monthly cycle one of these secondary follicles becomes dominant and develops further under the influence of FSH, LH and oestrogen.

Pre-Ovulatory Stage

The LH surge induces this stage and meiosis I is now complete. Two haploid cells are formed within the follicle, but they are of unequal size. One of the daughter cells receives far less cytoplasm than the other and forms the first polar body, which will not go on to form an ovum. The other haploid cell is known as the secondary oocyte. Both daughter cells then undergo meiosis II, the first polar body will replicated to give two polar bodies but the secondary oocyte arrests in metaphase of meiosis II, 3 hours prior to ovulation.

Ovulation

The follicle has grown in size and is now mature – it is called a Graafian follicle. The LH surge increases collagenase activity so that the follicular wall is weakened, this combined with muscular contractions of the ovarian wall result in the ovum being released from the ovary and being taken up into the fallopian tube via the fimbriae (finger-like projections of the fallopian tube).

Menstrual Cycle

Menstruation is the shedding of the lining of the uterus (endometrium) accompanied by bleeding. It occurs in approximately monthly cycles throughout a woman's reproductive life, except during pregnancy. Menstruation starts during puberty (at menarche) and stops permanently at menopause.

By definition, the menstrual cycle begins with the first day of bleeding, which is counted as day 1. The cycle ends just before the next menstrual period. Menstrual cycles normally range from about 25 to 36 days. Only 10 to 15% of women have cycles that are exactly 28 days. Also, in at least 20% of women, cycles are irregular. That is, they are longer or shorter than the normal range. Usually, the cycles vary the most and the intervals between periods are longest in the years immediately after menstruation starts (menarche) and before menopause.

Menstrual bleeding lasts 3 to 7 days, averaging 5 days. Blood loss during a cycle usually ranges from 1/2 to 2 1/2 ounces. A sanitary pad or tampon, depending on the type, can hold up to an ounce of blood. Menstrual blood, unlike blood resulting from an injury, usually does not clot unless the bleeding is very heavy.

The menstrual cycle is regulated by hormones. Luteinizing hormone and follicle-stimulating hormone, which are produced by the pituitary gland, promote ovulation and stimulate the ovaries to produce estrogen and progesterone. Estrogen and progesteronestimulate the uterus and breasts to prepare for possible fertilization.

The menstrual cycle has three phases:

- Follicular (before release of the egg),
- Ovulatory (egg release),
- Luteal (after egg release).

Follicular Phase

The follicular phase begins on the first day of menstrual bleeding (day 1). But the main event in this phase is the development of follicles in the ovaries.

At the beginning of the follicular phase, the lining of the uterus (endometrium) is thick with fluids and nutrients designed to nourish an embryo. If no egg has been fertilized, estrogen and progesterone levels are low. As a result, the top layers of the endometrium are shed, and menstrual bleeding occurs.

About this time, the pituitary gland slightly increases its production of follicle-stimulating hormone. This hormone then stimulates the growth of 3 to 30 follicles. Each follicle contains an egg. Later in the phase, as the level of this hormone decreases, only one of these follicles (called the dominant follicle) continues to grow. It soon begins to produce estrogen, and the other stimulated follicles begin to break down. The increasing estrogen also begins to prepare the uterus and stimulates the luteinizing hormone surge.

On average, the follicular phase lasts about 13 or 14 days. Of the three phases, this phase varies the most in length. It tends to become shorter near menopause. This phase ends when the level of luteinizing hormone increases dramatically (surges). The surge results in release of the egg (ovulation) and marks the beginning of the next phase.

Ovulatory Phase

The ovulatory phase begins when the level of luteinizing hormone surges. Luteinizing hormone stimulates the dominant follicle to bulge from the surface of the ovary and finally rupture, releasing the egg. The level of follicle-stimulating hormone increases to a lesser degree. The function of the increase in follicle-stimulating hormone is not understood.

The ovulatory phase usually lasts 16 to 32 hours. It ends when the egg is released, about 10 to 12 hours after the surge in the level of luteinizing hormone. The egg can be fertilized for only up to about 12 hours after its release.

The surge in luteinizing hormone can be detected by measuring the level of this hormone in urine. This measurement can be used to determine when women are fertile. Fertilization is more likely when sperm are present in the reproductive tract before the egg is released. Most pregnancies occur when intercourse occurs within 3 days before ovulation.

Around the time of ovulation, some women feel a dull pain on one side of the lower abdomen. This pain is known as mittelschmerz (literally, middle pain). The pain may last for a few minutes to a few hours. The pain is usually felt on the same side as the ovary that released the egg, but the precise cause of the pain is unknown. The pain may precede or follow the rupture of the follicle and may not occur in all cycles.

Egg release does not alternate between the two ovaries and appears to be random. If one ovary is removed, the remaining ovary releases an egg every month.

Luteal Phase

The luteal phase begins after ovulation. It lasts about 14 days (unless fertilization occurs) and ends just before a menstrual period.

In this phase, the ruptured follicle closes after releasing the egg and forms a structure called a corpus luteum, which produces increasing quantities of progesterone. The progesterone produced by the corpus luteum does the following:

- Prepares the uterus in case an embryo is implanted.
- Causes the endometrium to thicken, filling with fluids and nutrients to nourish a potential embryo.

- Causes the mucus in the cervix to thicken, so that sperm or bacteria are less likely to enter the uterus.
- Causes body temperature to increase slightly during the luteal phase and remain elevated until a menstrual period begins (this increase in temperature can be used to estimate whether ovulation has occurred).

During most of the luteal phase, the estrogen level is high. Estrogen also stimulates the endometrium to thicken.

The increase in estrogen and progesterone levels causes milk ducts in the breasts to widen (dilate). As a result, the breasts may swell and become tender.

If the egg is not fertilized or if the fertilized egg does not implant, the corpus luteum degenerates after 14 days, levels of estrogenand progesterone decrease, and a new menstrual cycle begins.

If the embryo is implanted, the cells around the developing embryo begin to produce a hormone called human chorionic gonadotropin. This hormone maintains the corpus luteum, which continues to produce progesterone, until the growing fetus can produce its own hormones.

Fertilization

The fusion of a haploid male gamete (sperm) and a haploid female gamete (ovum) to form a diploid zygote is called fertilization.

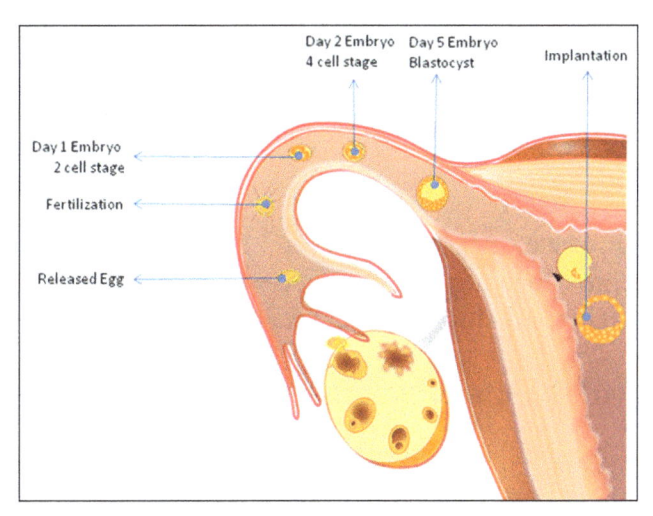

The idea of fertilization was known to Leeuwenhoek in 1683.

Site of Fertilization

In human beings, fertilization takes place mostly in the ampullary- isthmic junction of the oviduct (Fallopian tube).

Arrival of Sperms

Male discharges semen into the female's vagina close to the cervix during coitus (copulation). This is called insemination. A single ejaculation of semen may contain 300 million sperms.

Movement of Sperms

From the vagina the sperms travel up the uterus but only a few thousand find their way into the openings of the fallopian tubes.

Primarily, contractions of the uterus and fallopian tubes assist in sperm movement but later on they move by their own motility. Sperms swim in the fluid medium at the rate of 1.5 to 3 mm per minute to reach the site. The leucocytes of the vaginal epithelium engulf millions of sperms.

Arrival of Secondary Oocyte

In human beings, the secondary oocyte is released from the mature Graafian follicle of an ovary (ovulation). The oocyte is received by the nearby Fallopian funnel and sent into the Fallopian tube by movements of fimbriae and their cilia. The secondary oocyte can be fertilized only within 24 hours after its release from the ovary.

The secondary oocyte is surrounded by numerous sperms but only one sperm succeeds in fertilizing the oocyte. Since the second meiotic division is in progress, so the sperm enters the secondary oocyte. Second meiotic division is completed by the entry of the sperm into the secondary oocyte. After this secondary oocyte is called ovum (egg).

Capacitation of Sperms

The sperms in the female's genital tract are made capable of fertilizing the egg by secretions of the female genital tract. These secretions of the female genital tract remove coating substances deposited on the surface of the sperms particularly those on the acrosome.

Thus, the receptor sites on the acrosome are exposed and sperm becomes active to penetrate the egg. This phenomenon of sperm activation in mammals is known as capacitation. It takes about 5 to 6 hours for capacitation.

The secretions of seminal vesicles, prostate gland and bulbourethral glands (Cowper's glands) in the semen contain nutrients which activate the sperms. The secretions of these glands also neutralise the acidity in the vagina. Alkaline medium makes the sperms more active.

Physical and Chemical Events of Fertilization

These Events Include the following Processes:

(i) Acrosomal Reaction

After ovulation, the secondary oocyte reaches the Fallopian tube (oviduct). The capacitated sperms undergo acrosomal reaction and release various chemicals contained in the acrosome. These chemicals are collectively called sperm lysins. Important sperm lysins are (i) hyaluronidase that acts on the ground substances of follicle cells, (ii) corona penetrating enzyme that dissolves corona radiata and (iii) zona lysine or acrosin that helps to digest the zona pellucida.

Optimum pH, Ca^{++}, Mg^{++} ions concentration and temperature are essential for acrosomal reaction. Ca^{++} plays major role in acrosomal reaction. In the absence of Ca^{++}, fertilization does not occur.

Due to acrosomal reaction, plasma membrane of the sperm fuses with the plasma membrane of the secondary oocyte so that the sperm contents enter the oocyte. Binding of the sperm to the secondary oocyte induces depolarization of the oocyte plasma membrane. Depolarization prevents polyspermy (entry of more than one sperm into the oocyte). It ensures monospermy (entry of one sperm into the oocyte).

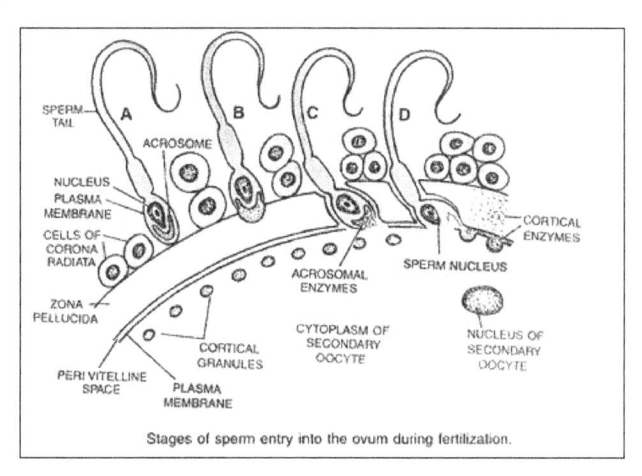

Stages of sperm entry into the ovum during fertilization.

(ii) Cortical Reaction

Just after the fusion of sperm and plasma membranes of oocyte, the secondary oocyte shows a cortical reaction. The cortical granules are present beneath the plasma membrane of the secondary oocyte.

These granules fuse with the plasma membrane of the oocyte and release their contents including cortical enzymes between the plasma membrane and the zona pellucida. These enzymes harden the zona pellucida which also prevents entry of additional sperms (polyspermy).

(iii) Sperm Entry

At the point of contact with the sperms, the secdndary oocyte forms a projection termed the cone of reception or fertilization cone which receives the sperm. The distal centriole of the sperm divides and forms two centrioles to generate the mitotic spindle formation for cell division. The mammalian secondary oocyte (egg) does not have centrioles of its own.

(iv) Karyogamy (Amphimixis)

Sperm entry stimulates the secondary oocyte to complete the suspended second meiotic division. This produces a haploid mature ovum and a second polar body. The head of the sperm which contains the nucleus separates from the middle piece and the tail and becomes the male pronucleus.

The second polar body and the sperm tail degenerate. The nucleus of the ovum is now called, the female pronucleus. The male and female pronuclei move towards each other. Their nuclear membranes disintegrate.

Mixing up of the chromosomes of a sperm and an ovum is known as karyogamy or amphimixis. The fertilized ovum (egg) is now called zygote. The zygote is diploid unicellular cell that has 46 chromosomes in humans. The mother is now said to be pregnant.

(v) Activation of Egg

Sperm entry stimulates metabolism in the zygote. As a result, the rates of cellular respiration and protein synthesis increase greatly. Besides activating the egg another role of sperm is to carry DNA to egg.

Significance of Fertilization

Fertilization has the following significances, (i) It restores the diploid number of chromosomes, characteristic of the species viz., 46 in human being, (ii) Fertilization initiates cleavage, (iii) It introduces the centrioles which are lacking in the mature egg. (iv) Fertilization results in determination of sex in the embryo, (v) It combines the characters of two parents. This introduces variations. (vi) Fertilization membrane developed after the entry of the sperm prevents the entry of other sperms into the ovum.

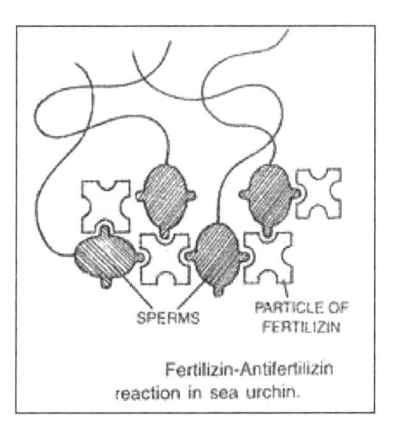

Fertilizin-Antifertilizin reaction in sea urchin.

Embryogenesis

Embryogenesis refers to the period of the first eight weeks of development after fertilization. It is an incredibly complicated process.

It's amazing that in eight weeks we're transforming from a single cell to an organism with a multi-level body plan. The circulatory, excretory, and neurologic systems all begin to develop during this stage. Luckily, like with many complex biological concepts, fertilization can be broken down into smaller, simpler ideas. The big idea of embryogenesis is going from a single cell to a ball of cells to a set of tubes.

The Beginning

- Step 1: A zygote is the single cell formed when an egg and a sperm cell fuse; the fusion is known as fertilization.
- Step 2: The first 12-to 24-hours after a zygote is formed are spent in cleavage – very rapid cell division.

The zygote's first priority is dividing to make lots of new cells, so it's first few days are spent in rapid mitotic division. With each round of division, it doubles in cell number, so the cell number is increasing at an exponential rate! This division is taking place so quickly that the cells don't have time to grow, so the 32 cell stage known as the morula is the same size as the zygote. At this point, the zona pellucida (a protective membrane of glycoproteins that had surrounded the egg cell) is still intact, which also limits how big it can grow.

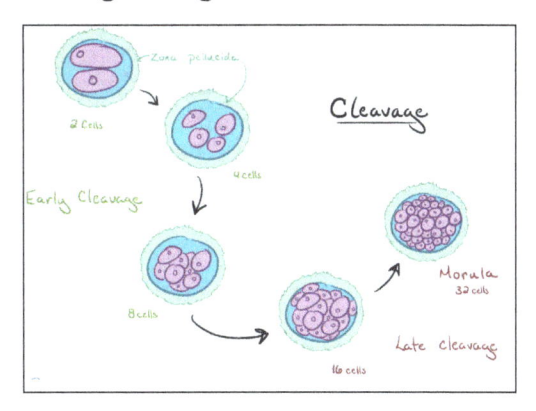

Blastulation and Cell Differentiation

- Step 3: During blastulation, the mass of cells forms a hollow ball.
- Step 4: Cells begin to differentiate, and form cavities.

Around day 4, cells continue to divide, but they also begin to differentiate and develop more specific forms and functions. When a cell differentiates, it moves down a certain path toward being a specific type of cell (e.g. an ear cell or a kidney cell), and this process (99% of the time) only goes in one direction. Two layers develop: an outer shell layer known as the trophoblast, and an inner collection of cells called the inner cell mass. Rather than being arranged in a solid sphere of cells, the inner cell mass is pushed off to one side of the sphere formed by the trophoblast. The rest of the fluid-filled cavity is called the blastocoel, and the whole setup resembles a snow globe. The outer trophoblast will develop into structures that help the growing embryo implant in the mother's uterus. The inner cell mass will continue to differentiate and parts of it will eventually become the embryo, so it is sometimes called the embryoblast (the suffix "blast" means "to make"). This is also the time when the zona pellucida begins to disappear, allowing the ball of cells, now called a blastocyst, to grow and change shape. In non-mammal animals, the term for this stage is "blastula".

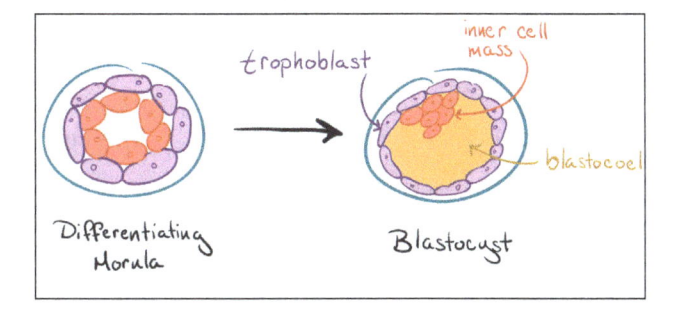

At this point, cells in the inner cell mass are pluripotent, meaning they can eventually turn into the cells of any body tissue (muscle, brain, bone, etc). During the second week, these cells differentiate further into the epiblast and the hypoblast, which are the two layers of the bilaminar disc. This disc is a flat slice across the developing sphere, and splits the environment into two cavities. The hypoblast is the layer facing the blastocoel, while the epiblast is on the other side. Let's imagine each of these layers as a flat balloon. The balloons expand to fill the space, and become the two new cavities: the primitive yolk sac on the side of the hypoblast and the amniotic cavity on the side of epiblast. The amniotic cavity will eventually surround the fetus.

The outermost layer of the sphere is the trophoblast. Inside the sphere are two spaces that are each lined by either the hypoblast or the epiblast. The point where the epiblast and hypoblast press up against each other is known as the bilaminar disc, and this disk is what splits the sphere to make the two cavities.

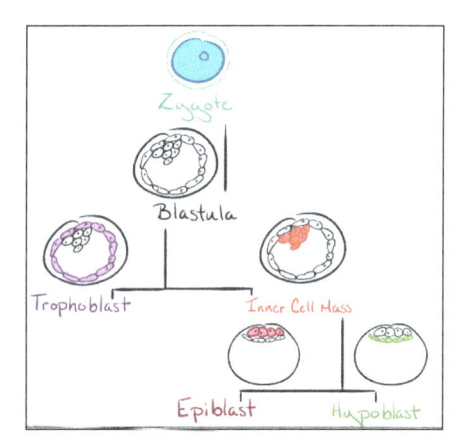

The hypoblast does not contribute to the embryo, so we will now turn our focus solely on the epiblast.

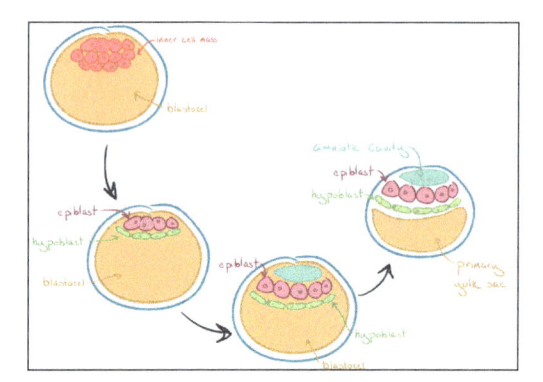

Making Tubes

- Step 5: During gastrulation the three germ layers form; the cell mass is now known as a gastrula.

- Step 5a: The primitive streak forms.

- Step 6: The notochord is formed.

Week 3 of development is the week of gastrulation. A germ layer is a layer of cells that will go on to form one of our organizational tubes. Our anatomy can really be boiled down to an inner tube (our digestive tract), and a series of tubes that wrap around it. The three germ layers that will translate into these tubes are the ectoderm, the mesoderm, and the endoderm.

component	percent of total cell weight
water	70
polysaccharides	2
Approximate chemical composition of a typical mammalian cell	

The first step of gastrulation is the formation of the primitive streak (~ day 16). Let's imagine the bilaminar disc as two tier cake. Imagine taking a knife and cutting into just the top layer (the epiblast) like you're going to cut a slice.

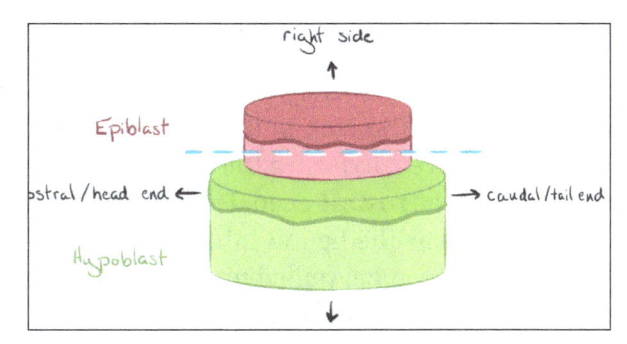

This cut is the primitive streak, and it cuts from the caudal (anus) end in toward the end that will eventually become the head (the rostral end). This streak determines the midline of the body, and separates the left and right sides. Like all deuterostomes, humans have bilateral symmetry, which means that there is a single across which we can split ourselves to make mirror images. What we are actually seeing when we look at a primitive streak are moving cells. They are going from the epiblast and moving down so they end up between the original epiblast layer and the hypoblast. I've always imagined the motion like water falling down a waterfall. The first layer to invaginate dives the deepest and ends up closest to the hypoblast – this is the endoderm. The next layers will become the mesoderm, and the cells of the epiblast that continue to border the amniotic cavity are the ectoderm. We now have three germ layers, all of which will contribute to the developing embryo.

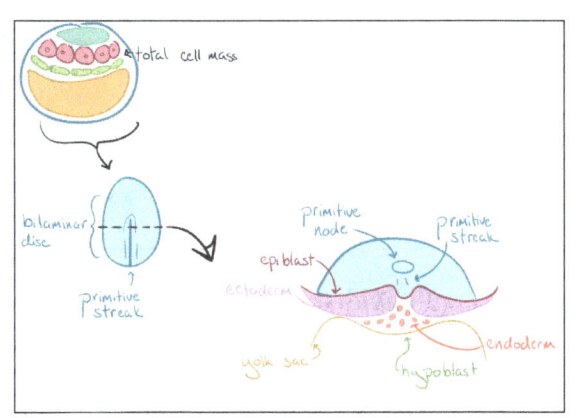

Directly beneath the primitive streak the mesoderm (the middle germ layer) forms a thin rod of cells known as the notochord. The notochord helps define the major axis of our bodies, and is important in inducing the next step of embryogenesis, when we finally start to make our tubes! The notochord is a defining feature of the Chordate phylum, and will eventually become our intervertebral discs.

Neurulation

- Step 6: Tubes form, making a neurula,
- Step 6a: The notochord induces the formation of the neural plate,
- Step 6b: The neural plate folds in on itself to make the neural tube and neural crest,
- Step 7: The mesoderm has five distinct categories.

All this and we still haven't made tubes! Now that we have successfully made the cell layers, we have to create the final 3D product. The first step in this rolling is the creation of the notochord. The notochord causes the ectoderm above it to form a thick flat plate of cells called the neural plate. The neural plate extends the length of the rostral-caudal axis. The neural plate then bends back on itself and seals itself into a tube known as the neural tube that fits underneath the ectoderm. The borders of where the neural plate had been get pulled under with it, and become the neural crest. The neural tube will become the brain and spinal cord.

The neural crest is sometimes called the fourth germ layer, because the cells that become the sympathetic and parasympathetic nervous systems, melanocytes, Schwann cells, even some of the bones and connective tissue of the face.

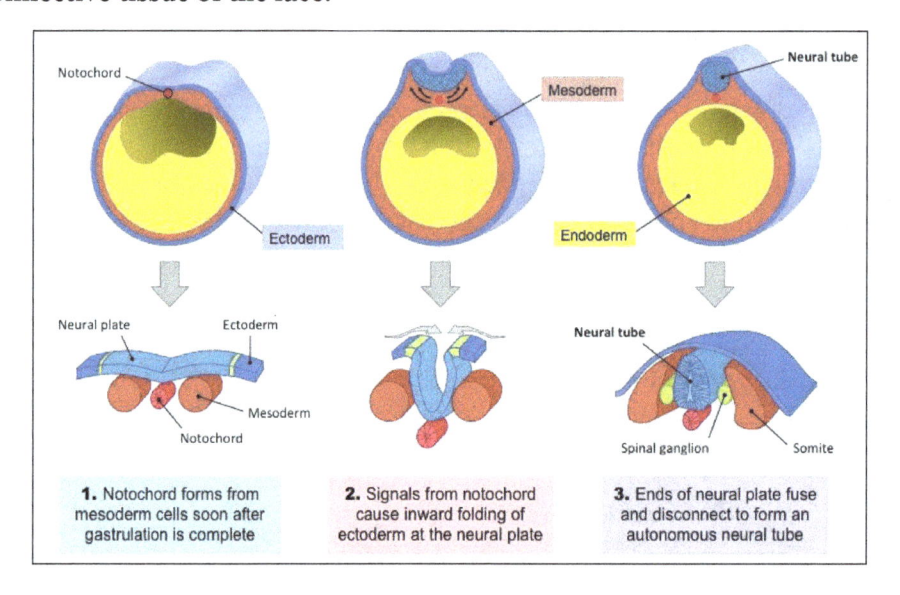

Meanwhile, the mesoderm can be subdivided into the axial, paraxial, intermediate, and lateral plate mesoderms. The notochord came from the axial mesoderm. The paraxial mesoderm will give rise to somites, which will differentiate into muscle, cartilage, bone, and dermis. Somite derivatives create a segmented body plan. The intermediate mesoderm is the origin of our urogenital system – our kidneys, gonads, adrenal glands, and the ducts that connect them. The lateral plate mesoderm will give rise to the heart (the first organ to develop), blood vessels, the body wall, and the muscle in our organs.

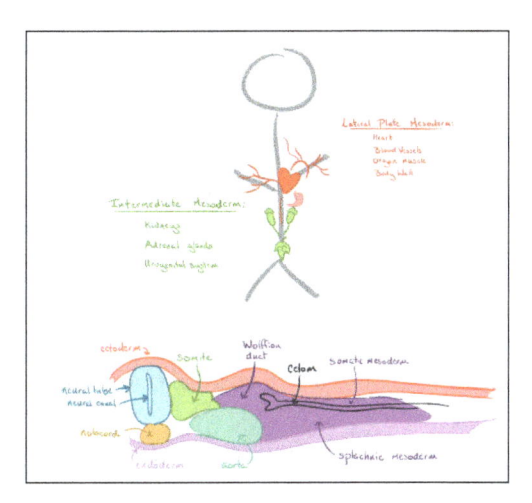

Also at the same time, the endoderm is rolling into a tube as well – the digestive tract. The digestive tract is subdivided into the foregut, midgut, and hindgut. Each subdivision has its own nerve and blood supply. Organs related to the GI tract actually start off as outpouchings of this tube. The foregut gives rise to the esophagus, stomach, part of the duodenum, and the respiratory bud, which will eventually develop into the lungs. The second half of the duodenum through to the transverse colon arise from the midgut. The remainder of the GI tract, including the rest of the transverse colon, the descending colon, the sigmoid colon, and the rectum are formed from the hindgut.

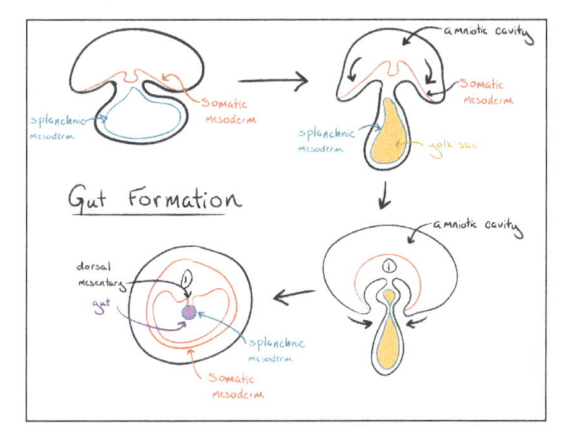

That's what is going in with each of the three layers. While this is happening, the mesodermal layers are circling around the endoderm, and the part of the ectoderm that will become the skin is circling around both of the other layers. Some tubes, likes the neural tube, are closing, while the gut tube is connecting to the ectoderm to form the mouth and the anus. By the time eight weeks have passed, all of our tubes are in order, the primitive heart has been beating for almost five weeks, and development is well on its way.

References

- Reproduction-biology, science: britannica.com, Retrieved 15 May, 2019
- Asexual-reproduction: biologydictionary.net, Retrieved 5 January, 2019
- Sexual-Reproduction: biologyreference.com, Retrieved April, 2019
- Sexual-reproduction, reproduction-in-animals, biology: toppr.com, Retrieved 25 February, 2019

- Sexual-reproduction, biology: lumenlearning.com, Retrieved 27 July, 2019
- Reproduction-in-Plants: biologyreference.com, Retrieved 8 June, 2019
- Sexual-reproduction-in-plant, reproduction-in-plants, science: toppr.com, Retrieved 18 August, 2019
- Double-fertilization-in-angiosperms, biology: byjus.com, Retrieved 2 March, 2019
- Reproduction-human-beings, how-do-organisms-reproduce, biology: toppr.com, Retrieved 27 February, 2019
- Human-reproductive-system, science: britannica.com, Retrieved 23 July, 2019
- Gametogenesis, embryology, reproductive-system: teachmephysiology.com, Retrieved 21 January, 2019
- Menstrual-cycle, biology-of-the-female-reproductive-system, women-s-health: msdmanuals.com, Retrieved 30 August, 2019
- Fertilization-notes-on-fertilization-in-humans, biology: yourarticlelibrary.com, Retrieved 3 March, 2019
- Human-embryogenesis, embryology, cells: khanacademy.org, Retrieved 26 May, 2019

Chapter 4
Inheritance and Variation

Biological inheritance is the process of passing the traits from parents to their offspring. The offspring cells and organisms inherit the genetic information of their parents either by asexual reproduction or sexual reproduction. All these diverse principles of inheritance, mutation and variation have been carefully analyzed in this chapter.

Biological Inheritance

Biological inheritance is the process by which an offspring cell or organism acquires or becomes predisposed to characteristics of its parent cell or organism. Through inheritance, variations exhibited by individuals can accumulate and cause a species to evolve. The study of biological inheritance is called genetics.

Categories of Biological Inheritance

The Description of a Mode of Biological Inheritance Consists of three main Categories:

1. Number of involved loci:
   - Monogenetic (also called 'simple')—one locus.
   - Oligogenetic—few loci.
   - Polygenetic—many loci.

2. Involved chromosomes:
   - Autosomal—Loci are not situated on a sex chromosome.
   - Gonosomal—Loci are situated on a sex chromosome.
   - X-chromosomal—Loci are situated on the X chromosome (the more common case).
   - Y-chromosomal—Loci are situated on the Y chromosome.
   - Mitochondrial—Loci are situated on the mitochondrial DNA.

3. Correlation genotype-phenotype:
   - Dominant.
   - Intermediate (also called 'co-dominant').
   - Recessive.

These three categories are part of every exact description of a mode of inheritance in the above order.

Specifications of Biological Inheritance

Additionally, more specifications may be added as follows:

1. Coincidental and environmental interactions:
   - Penetrance.
   - Incomplete (percentual number).
   - Invariable.
   - Complete.
   - Expressivity.
   - Variable.
   - Maternal or paternal imprinting phenomena.
   - Heritability (in polygenetic and sometimes also in oligogenetic modes of inheritance).

2. Sex-linked interactions:
   - Sex-linked inheritance (gonosomal loci).
   - Sex-limited phenotype expression (ex. Cryptorchism).
   - Inheritance through the maternal line (in case of mitochondrial DNA loci).
   - Inheritance through the paternal line (in case of Y-chromosomal loci).

3. Locus-locus-interactions:
   - Epistasis with other loci (ex. Over-dominance).
   - Gene coupling with other loci.
   - Homozygotes lethal factors.
   - Semi-lethal factors.

Determination and description of a mode of inheritance is primarily achieved through statistical analysis of pedigree data. In case the involved loci are known, methods of molecular genetics can also be employed. Mendelian inheritance (or Mendelian genetics or Mendelism) is a set of primary tenets relating to the transmission of hereditary characteristics from parent organisms to their children.

Mendel's Experiment

Mendel began a decade-long research project to investigate patterns of inheritance. Although he began his research using mice, he later switched to honeybees and plants, ultimately settling on garden peas as his primary model system. A model system is an organism that makes it easy for a researcher to investigate a particular scientific question, such as how traits are inherited. By studying a model system, researchers can learn general principles that apply to other, harder-to-study organisms or biological systems, such as humans.

Mendel studied the inheritance of seven different features in peas, including height, flower color, seed color, and seed shape. To do so, he first established pea lines with two different forms of a feature, such as tall vs. short height. He grew these lines for generations until they were pure-breeding (always produced offspring identical to the parent), then bred them to each other and observed how the traits were inherited.

In addition to recording how the plants in each generation looked, Mendel counted the exact number of plants that showed each trait. Strikingly, he found very similar patterns of inheritance for all seven features he studied:

- One form of a feature, such as tall, always concealed the other form, such as short, in the first generation after the cross. Mendel called the visible form the dominant trait and the hidden form the recessive trait.

- In the second generation, after plants were allowed to self-fertilize (pollinate themselves), the hidden form of the trait reappeared in a minority of the plants. Specifically, there were always about 3 plants that showed the dominant trait (e.g., tall) for every 1 plant that showed the recessive trait (e.g., short), making a 3:1 ratio.

- Mendel also found that the features were inherited independently: one feature, such as plant height, did not influence inheritance of other features, such as flower color or seed shape.

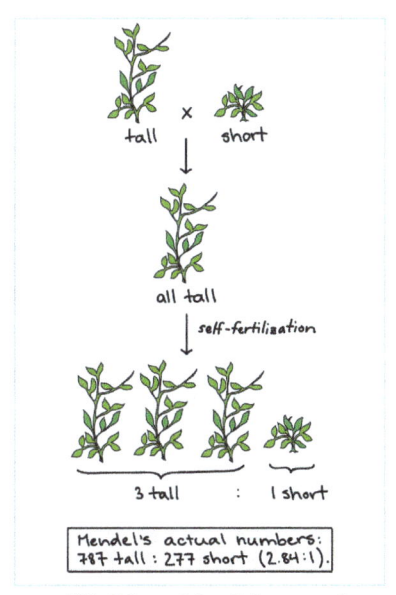

Image modified from Mendel seven characters.

In 1865, Mendel presented the results of his experiments with nearly 30,000 pea plants to the local Natural History Society. Based on the patterns he observed, the counting data he collected, and a mathematical analysis of his results, Mendel proposed a model of inheritance in which:

- Characteristics such as flower color, plant height, and seed shape were controlled by pairs of heritable factors that came in different versions.

- One version of a factor (the dominant form) could mask the presence of another version (the recessive form).

- The two paired factors separated during gamete production, such that each gamete (sperm or egg) randomly received just one factor.

- The factors controlling different characteristics were inherited independently of one another.

Mendel's Model System: The Pea Plant

Mendel carried out his key experiments using the garden pea, Pisum sativum, as a model system. Pea plants make a convenient system for studies of inheritance, and they are still studied by some geneticists today.

Useful features of peas include their rapid life cycle and the production of lots and lots of seeds. Pea plants also typically self-fertilize, meaning that the same plant makes both the sperm and the egg that come together in fertilization. Mendel took advantage of this property to produce true-breeding pea lines: he self-fertilized and selected peas for many generations until he got lines that consistently made offspring identical to the parent (e.g., always short).

Pea plants are also easy to cross, or mate in a controlled way. This is done by transferring pollen from the anthers (male parts) of a pea plant of one variety to the carpel (female part) of a mature pea plant of a different variety. To prevent the receiving plant from self-fertilizing, Mendel painstakingly removed all of the immature anthers from the plant's flowers before the cross.

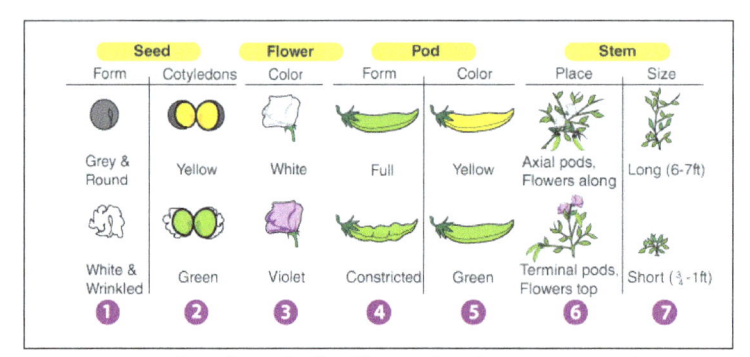

Image based on similar illustration from Reece et al.

Because peas were so easy to work with and prolific in seed production, Mendel could perform many crosses and examine many individual plants, making sure that his results were consistent (not just a fluke) and accurate (based on many data points).

Mendel's Experimental Setup

Once Mendel had established true-breeding pea lines with different traits for one or more features of interest (such as tall vs. short height), he began to investigate how the traits were inherited by carrying out a series of crosses.

First, he crossed one true-breeding parent to another. The plants used in this initial cross are called the P generation, or parental generation.

Mendel collected the seeds from the P generation cross and grew them up. These offspring were called the F_1 generation, short for first filial generation.

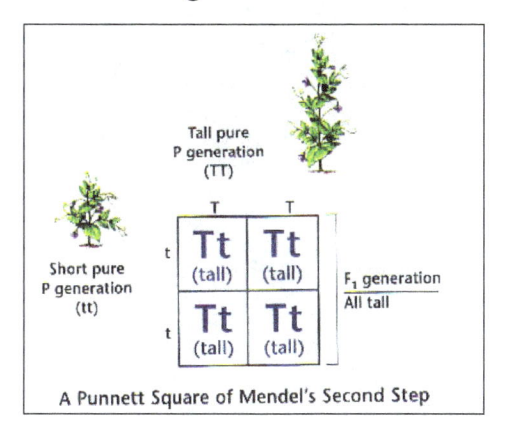

A Punnett Square of Mendel's Second Step

Once Mendel examined the F_1 plants and recorded their traits, he let them self-fertilize naturally, producing lots of seeds. He then collected and grew the seeds from the F_1 plants to produce an F_2 generation, or second filial generation. Again, he carefully examined the plants and recorded their traits.

Mendel's experiments extended beyond the F_2 generation to F_3 and F_4 and later generations, but his model of inheritance was based mostly on the first three generations (P F_1 and F_2).

Mendel didn't just record what his plants looked like in each generation (e.g., tall vs. short). Instead, he counted exactly how many plants with each trait were present.

Laws of Inheritance

Mendel generalized the results of his pea-plant experiments into four postulates, some of which are sometimes called "laws," that describe the basis of dominant and recessive inheritance in diploid organisms. these laws summarize the basics of classical genetics.

Pairs of Unit Factors or Genes

Mendel proposed first that paired unit factors of heredity were transmitted faithfully from generation to generation by the dissociation and reassociation of paired factors during gametogenesis and fertilization, respectively. After he crossed peas with contrasting traits and found that the recessive trait resurfaced in the F_2 generation, Mendel deduced that hereditary factors must be inherited as discrete units. This finding contradicted the belief at that time that parental traits were blended in the off spring.

Alleles can be Dominant or Recessive

Mendel's law of dominance states that in a heterozygote, one trait will conceal the presence of another trait for the same characteristic. Rather than both alleles contributing to a phenotype, the dominant allele will be expressed exclusively. The recessive allele will remain "latent" but will be transmitted to offspring by the same manner in which the dominant allele is transmitted. The recessive trait will only be expressed by offspring that have two copies of this allele, and these offspring will breed true when self-crossed.

The child in the photo expresses albinism, a recessive trait.

Since Mendel's experiments with pea plants, other researchers have found that the law of dominance does not always hold true. Instead, several different patterns of inheritance have been found to exist.

Equal Segregation of Alleles

Observing that true-breeding pea plants with contrasting traits gave rise to F_1 generations that all expressed the dominant trait and F_2 generations that expressed the dominant and recessive traits in a 3:1 ratio, Mendel proposed the law of segregation. This law states that paired unit factors (genes) must segregate equally into gametes such that offspring have an equal likelihood of inheriting either factor. For the F_2 generation of a monohybrid cross, the following three possible combinations of genotypes could result: homozygous dominant, heterozygous, or homozygous recessive. Because heterozygotes could arise from two different pathways (receiving one dominant and one recessive allele from either parent), and because heterozygotes and homozygous dominant individuals are phenotypically identical, the law supports Mendel's observed 3:1 phenotypic ratio. The equal segregation of alleles is the reason we can apply the Punnett square to accurately predict the offspring of parents with known genotypes. The physical basis of Mendel's law of segregation is the first division of meiosis, in which the homologous chromosomes with their different versions of each gene are segregated into daughter nuclei. The role of the meiotic segregation of chromosomes in sexual reproduction was not understood by the scientific community during Mendel's lifetime.

Independent Assortment

Mendel's law of independent assortment states that genes do not influence each other with regard to the sorting of alleles into gametes, and every possible combination of alleles for every gene is equally likely to occur. The independent assortment of genes can be illustrated by the dihybrid cross, a cross between two true-breeding parents that express different traits for two characteristics. Consider the characteristics of seed color and seed texture for two pea plants, one that has green, wrinkled seeds (yyrr) and another that has yellow, round seeds (YYRR). Because each parent is homozygous, the law of segregation indicates that the gametes for the green/wrinkled plant all are yr, and the gametes for the yellow/round plant are all YR. Therefore, the F_1 generation of offspring all are YyRr.

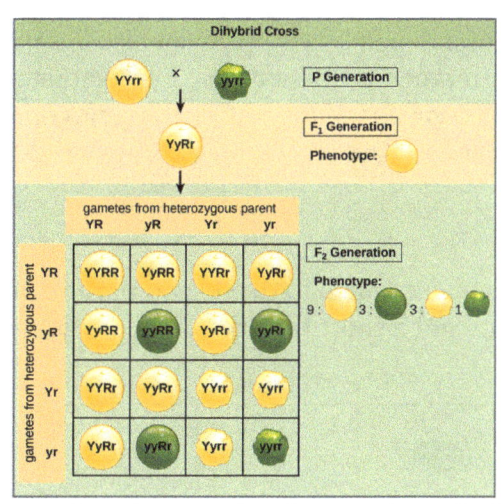

This dihybrid cross of pea plants involves the genes for seed color and texture.

In pea plants, purple flowers (P) are dominant to white flowers (p) and yellow peas (Y) are dominant to green peas (y). What are the possible genotypes and phenotypes for a cross between PpYY and ppYy pea plants? How many squares do you need to do a Punnett square analysis of this cross?

For the F_2 generation, the law of segregation requires that each gamete receive either an R allele or an r allele along with either a Y allele or a y allele. The law of independent assortment states that a gamete into which an r allele sorted would be equally likely to contain either a Y allele or a y allele. Thus, there are four equally likely gametes that can be formed when the YyRr heterozygote is self-crossed, as follows: YR, Yr, yR, and yr. Arranging these gametes along the top and left of a 4 × 4 Punnett square gives us 16 equally likely genotypic combinations. From these genotypes, we infer a phenotypic ratio of 9 round/yellow:3 round/green:3 wrinkled/yellow:1 wrinkled/green. These are the offspring ratios we would expect, assuming we performed the crosses with a large enough sample size.

Because of independent assortment and dominance, the 9:3:3:1 dihybrid phenotypic ratio can be collapsed into two 3:1 ratios, characteristic of any monohybrid cross that follows a dominant and recessive pattern. Ignoring seed color and considering only seed texture in the above dihybrid cross, we would expect that three quarters of the F2 generation offspring would be round, and one quarter would be wrinkled. Similarly, isolating only seed color, we would assume that three quarters of the F2 offspring would be yellow and one quarter would be green. The sorting of alleles for texture and color are independent events, so we can apply the product rule. Therefore, the proportion of round and yellow F2 offspring is expected to be (3/4) × (3/4) = 9/16, and the proportion of wrinkled and green offspring is expected to be (1/4) × (1/4) = 1/16. These proportions are identical to those obtained using a Punnett square. Round, green and wrinkled, yellow offspring can also be calculated using the product rule, as each of these genotypes includes one dominant and one recessive phenotype. Therefore, the proportion of each is calculated as (3/4) × (1/4) = 3/16.

The law of independent assortment also indicates that a cross between yellow, wrinkled (YYrr) and green, round (yyRR) parents would yield the same F1 and F2 offspring as in the YYRR x yyrr cross.

The physical basis for the law of independent assortment also lies in meiosis I, in which the different homologous pairs line up in random orientations. Each gamete can contain any combination of paternal and maternal chromosomes (and therefore the genes on them) because the orientation of tetrads on the metaphase plane is random.

Monohybrid Cross

It is the cross between two pea plants which have one pair of contrasting characters. For Example, a cross between a tall pea plant and a short (dwarf) plant. The following diagram explains this in detail.

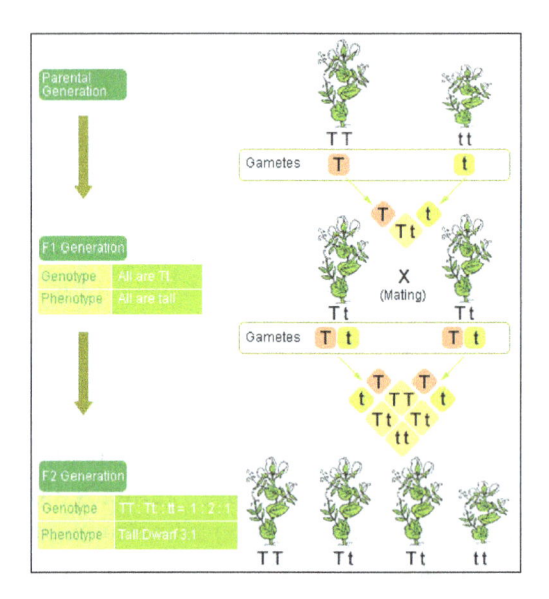

- In the first generation (F1), the progeny were tall. There was no medium height plant.

- In the second generation (F2), 1/4th of the offspring were short and 3/4 were tall.

- The Phenotypic ratio in F2 is 3: 1 (3 tall: 1 short).

- The Genotypic ratio in F2 – 1: 2: 1 – (TT: Tt: tt).

- For a plant to be tall, a single copy of "T" is enough. But if a plant has to be short, both the copies should be "t".

- Characters like 'T' are the dominant traits as they are expressed and 't' are recessive traits as they remain suppressed.

Dihybrid Cross

It is the cross between two plants which have two pairs of contrasting characters. This takes into consideration alternative traits of two different characters. For example, a cross between one pea plant with round and green seeds and the other pea plant having wrinkled and yellow seeds. The following diagram explains the dihybrid cross in detail.

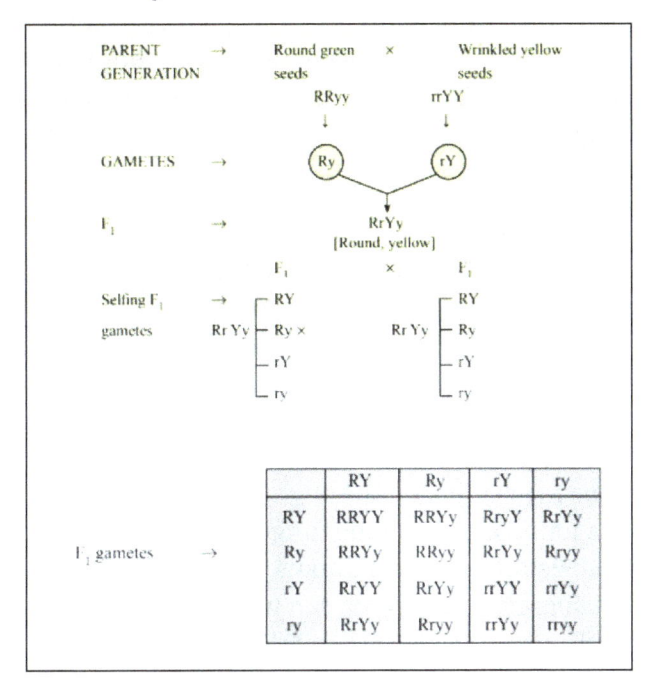

```
PARENT          →   Round green    ×    Wrinkled yellow
GENERATION          seeds               seeds
                    RRyy                rrYY
                     ↓                   ↓
GAMETES         →   (Ry)                (rY)
                              ↓
F1              →           RrYy
                        [Round, yellow]
                    F1      ×      F1

Selfing F1      →    ┌ RY              ┌ RY
 gametes       RrYy │ Ry    ×   RrYy  │ Ry
                    │ rY               │ rY
                    └ ry               └ ry
```

	RY	Ry	rY	ry
RY	RRYY	RRYy	RrYY	RrYy
Ry	RRYy	RRyy	RrYy	Rryy
rY	RrYY	RrYy	rrYY	rrYy
ry	RrYy	Rryy	rrYy	rryy

(F1 gametes →)

- The F1 generation is 100% hybrid. When RRyy crosses with rrYY, all were Rr Yy with round and yellow seeds in the first generation. The Round and Yellow seeds are the dominant characters.

- In F2, the phenotype ratio is 9:3:3:1. The genotype ratio is a very complex one.

- This shows that the genes are inherited independently of each other.

Heredity

Heredity is the sum of all biological processes by which particular characteristics are transmitted from parents to their offspring. The concept of heredity encompasses two seemingly paradoxical observations about organisms: the constancy of a species from generation to generation and the variation among individuals within a species. Constancy and variation are actually two sides of the same coin, as becomes clear in the study of genetics. Both aspects of heredity can be explained by genes, the functional units of heritable material that are found within all living cells. Every member of a species has a set of genes specific to that species. It is this set of genes that provides the constancy of the species. Among individuals within a species, however, variations can occur in the form each gene takes, providing the genetic basis for the fact that no two individuals (except identical twins) have exactly the same traits.

The set of genes that an offspring inherits from both parents, a combination of the genetic material of each, is called the organism's genotype. The genotype is contrasted to the phenotype, which is the organism's outward appearance and the developmental outcome of its genes. The phenotype includes an organism's bodily structures, physiological processes, and behaviours. Although the genotype determines the broad limits of the features an organism can develop, the features that actually develop, i.e., the phenotype, depend on complex interactions between genes and their environment. The genotype remains constant throughout an organism's lifetime; however, because the organism's internal and external environments change continuously, so does its phenotype. In conducting genetic studies, it is crucial to discover the degree to which the observable trait is attributable to the pattern of genes in the cells and to what extent it arises from environmental influence.

Because genes are integral to the explanation of hereditary observations, genetics also can be defined as the study of genes. Discoveries into the nature of genes have shown that genes are important determinants of all aspects of an organism's makeup. For this reason, most areas of biological research now have a genetic component, and the study of genetics has a position of central importance in biology. Genetic research also has demonstrated that virtually all organisms on this planet have similar genetic systems, with genes that are built on the same chemical principle and that function according to similar mechanisms. Although species differ in the sets of genes they contain, many similar genes are found across a wide range of species. For example, a large proportion of genes in baker's yeast are also present in humans. This similarity in genetic makeup between organisms that have such disparate phenotypes can be explained by the evolutionary relatedness of virtually all life-forms on Earth. This genetic unity has radically reshaped the understanding of the relationship between humans and all other organisms. Genetics also has had a profound impact on human affairs. Throughout history humans have created or improved many different medicines, foods, and textiles by subjecting plants, animals, and microbes to the ancient techniques of selective breeding and to the modern methods of recombinant DNA technology. In recent years medical researchers have begun to discover the role that genes play in disease. The significance of genetics only promises to become greater as the structure and function of more and more human genes are characterized.

The Physical basis of Heredity

When Gregor Mendel formulated his laws of heredity, he postulated a particulate nature for the units of inheritance. What exactly these particles were he did not know. Today scientists understand not only the physical location of hereditary units (i.e., the genes) but their molecular composition as well. The unraveling of the physical basis of heredity makes up one of the most fascinating chapters in the history of biology.

Chromosomes and Genes

Each individual in a sexually reproducing species inherits two alleles for each gene, one from each parent. Furthermore, when such an individual forms sex cells, each of the resultant gametes receives one member of each allelic pair. The formation of gametes occurs through a process of cell division called meiosis. When gametes unite in fertilization, the double dose of hereditary material is restored, and a new individual is created. This individual, consisting at first of only one cell, grows via mitosis, a process of repeated cell divisions. Mitosis differs from meiosis in that each daughter cell receives a full copy of all the hereditary material found in the parent cell.

Inheritance and Variation

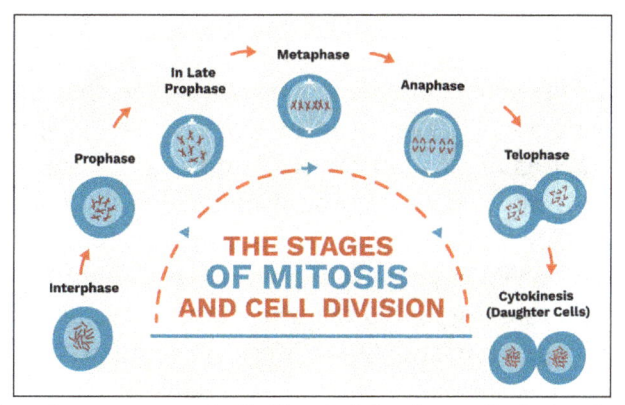

Mitosis: One cell gives rise to two genetically identical daughter cells during the process of mitosis.

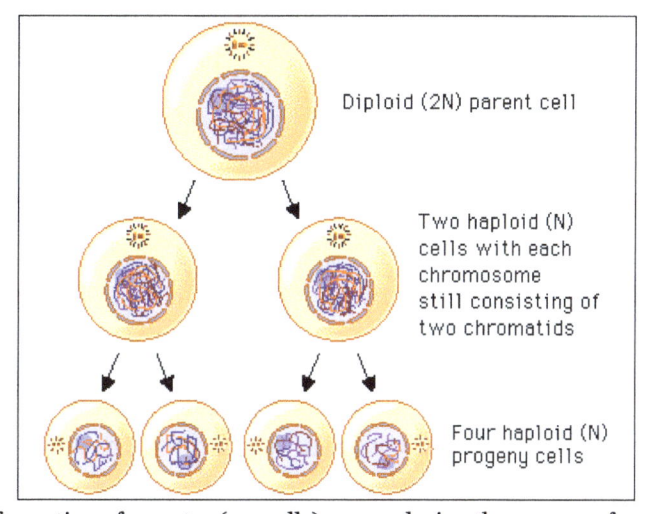

The formation of gametes (sex cells) occurs during the process of meiosis.

It is apparent that the genes must physically reside in cellular structures that meet two criteria. First, these structures must be replicated and passed on to each generation of daughter cells during mitosis. Second, they must be organized into homologous pairs, one member of which is parceled out to each gamete formed during meiosis.

As early as 1848, biologists had observed that cell nuclei resolve themselves into small rodlike bodies during mitosis; later these structures were found to absorb certain dyes and so came to be called chromosomes (coloured bodies). During the early years of the 20th century, cellular studies using ordinary light microscopes clarified the behaviour of chromosomes during mitosis and meiosis, which led to the conclusion that chromosomes are the carriers of genes.

The Behavior of Chromosomes during Cell Division

During Mitosis

When the chromosomes condense during cell division, they have already undergone replication. Each chromosome thus consists of two identical replicas, called chromatids, joined at a point called the centromere. During mitosis the sister chromatids separate, one going to each daughter cell. Chromosomes thus meet the first criterion for being the repository of genes: they are replicated, and a full copy is passed to each daughter cell during mitosis.

During Meiosis

It was the behaviour of chromosomes during meiosis, however, that provided the strongest evidence for their being the carriers of genes. In 1902 American scientist Walter S. Sutton reported on his observations of the action of chromosomes during sperm formation in grasshoppers. Sutton had observed that, during meiosis, each chromosome (consisting of two chromatids) becomes paired with a physically similar chromosome. These homologous chromosomes separate during meiosis, with one member of each pair going to a different cell. Assuming that one member of each homologous pair was of maternal origin and the other was paternally derived, here was an event that fulfilled the behaviour of genes postulated in Mendel's first law.

It is now known that the number of chromosomes within the nucleus is usually constant in all individuals of a given species—for example, 46 in the human, 40 in the house mouse, 8 in the vinegar fly (Drosophila melanogaster; sometimes called fruit fly), 20 in corn (maize), 24 in the tomato, and 48 in the potato. In sexually reproducing organisms, this number is called the diploid number of chromosomes, as it represents the double dose of chromosomes received from two parents. The nucleus of a gamete, however, contains half this number of chromosomes, or the haploid number. Thus, a human gamete contains 23 chromosomes, while a Drosophila gamete contains four. Meiosis produces the haploid gametes.

The essential features of meiosis are shown in the diagram. For the sake of simplicity, the diploid parent cell is shown to contain a single pair of homologous chromosomes, one member of which is represented in blue (from the father) and the other in red (from the mother). At the leptotene stage the chromosomes appear as long, thin threads. At pachytene they pair, the corresponding portions of the two chromosomes lying side by side. The chromosomes then duplicate and contract into paired chromatids. At this stage the pair of chromosomes is known as a tetrad, as it consists of four chromatids. Also at this stage an extremely important event occurs: portions of the maternal and paternal chromosomes are exchanged. This exchange process, called crossing over, results in chromatids that include both paternal and maternal genes and consequently introduces new genetic combinations. The first meiotic division separates the chromosomal tetrads, with the paternal chromosome (whose chromatids now contain some maternal genes) going to one cell and the maternal chromosome (containing some paternal genes) going to another cell. During the second meiotic division the chromatids separate. The original diploid cell has thus given rise to four haploid gametes (only two of which are shown in the diagram). Not only has a reduction in chromosome number occurred, but the resulting single member of each homologous chromosome pair may be a new combination (through crossing over) of genes present in the original diploid cell.

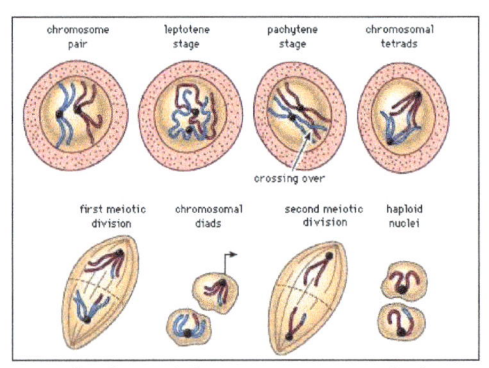

Behaviour of chromosomes at meiosis.

Suppose that the red chromosome shown in the diagram carries the gene for albinism, and the blue chromosome carries the gene for dark pigmentation. It is evident that the two gene alleles will undergo segregation at meiosis and that one-half of the gametes formed will contain the albino gene and the other half the pigmentation gene. Following the scheme in the diagram, random combination of the gametes with the albino gene and the pigmentation gene will give two kinds of homozygotes and one kind of heterozygote in a ratio of 1 : 1 : 2. Mendel's law of segregation is thus the outcome of chromosome behaviour at meiosis. The same is true of the second law, that of independent assortment.

Consider the inheritance of two pairs of genes, such as Mendel's factors for seed coloration and seed surface in peas; these genes are located on different pairs of chromosomes. Since maternal and paternal members of different chromosome pairs are assorted independently, so are the genes they contain. This explains, in part, the genetic variety seen among the progeny of the same pair of parents. As stated above, humans have 46 chromosomes in the body cells and in the cells (oogonia and spermatogonia) from which the sex cells arise. At meiosis these 46 chromosomes form 23 pairs, one of the chromosomes of each pair being of maternal and the other of paternal origin. Independent assortment is, then, capable of producing 2^{23}, or 8,388,608, kinds of sex cells with different combinations of the grandmaternal and grandpaternal chromosomes. Since each parent has the potentiality of producing 2^{23} kinds of sex cells, the total number of possible combinations of the grandparental chromosomes is $2^{23} \times 2^{23} = 2^{46}$. The population of the world is now more than 6 billion persons, or approximately 2^{32} persons. It is therefore certain that only a tiny fraction of the potentially possible chromosome and gene combinations can ever be realized. Yet even 2^{46} is an underestimate of the variety potentially possible. The grandmaternal and grandpaternal members of the chromosome pairs are not indivisible units. Each chromosome carries many genes, and the chromosome pairs exchange segments at meiosis through the process of crossing over. This is evidence that the genes rather than the chromosomes are the units of Mendelian segregation.

Linkage of Traits

Simple Linkage

As pointed out above, the random assortment of the maternal and paternal chromosomes at meiosis is the physical basis of the independent assortment of genes and of the traits they control. This is the basis of the second law of Mendel. The number of the genes in a sex cell is, however, much greater than that of the chromosomes. When two or more genes are borne on the same chromosome, these genes may not be assorted independently; such genes are said to be linked. When a Drosophila fly homozygous for a normal gray body and long wings is crossed with one having a black body and vestigial wings, the F_1 consists of hybrid gray, long-winged flies. Gray body (B) is evidently dominant over black body (b), and long wing (V) is dominant over vestigial wing (v). Now consider a backcross of the heterozygous F1 males to double-recessive black-vestigial females (bbvv). Independent assortment would be expected to give in the progeny of the backcross the following: 1 gray-long : 1 gray-vestigial : 1 black-long : 1 black-vestigial. In reality, only gray-long and black-vestigial flies are produced, in approximately equal numbers; the genes remain linked in the same combinations in which they were found in the parents. The backcross of the heterozygous F1 females to double-recessive males gives a somewhat different result: 42 percent each of gray-long and black-vestigial flies and about 8 percent each of black-long and gray-vestigial classes. In sum,

84 percent of the progeny have the parental combinations of traits, and 16 percent have the traits recombined. The interpretation of these results given in 1911 by the American geneticist Thomas Hunt Morgan laid the foundation of the theory of linear arrangement of genes in the chromosomes.

Traits that exhibit linkage in experimental crosses (such as black body and vestigial wings) are determined by genes located in the same chromosome. As more and more genes became known in Drosophila, they fell neatly into four linkage groups corresponding to the four pairs of the chromosomes this species possesses. One linkage group consists of sex-linked genes, located in the X chromosome; of the three remaining linkage groups, two have many more genes than the remaining one, which corresponds to the presence of two pairs of large chromosomes and one pair of tiny dotlike chromosomes. The numbers of linkage groups in other organisms are equal to or smaller than the numbers of the chromosomes in the sex cells—e.g., 10 linkage groups and 10 chromosomes in corn, 19 linkage groups and 20 chromosomes in the house mouse, and 23 linkage groups and 23 chromosomes in the human.

The linkage of the genes black and vestigial in Drosophila is complete in heterozygous males, while in the progeny of females there appear about 17 percent of recombination classes. With very rare exceptions, the linkage of all genes belonging to the same linkage group is complete in Drosophila males, while in the females different pairs of genes exhibit all degrees of linkage from complete (no recombination) to 50 percent (random assortment). Morgan's inference was that the degree of linkage depends on physical distance between the genes in the chromosome: the closer the genes, the tighter the linkage and vice versa. Furthermore, Morgan perceived that the chiasmata (crosses that occur in meiotic chromosomes) indicate the mechanism underlying the phenomena of linkage and crossing over. As shown schematically in the diagram of chromosomes at meiosis, the maternal and paternal chromosomes (represented in blue and red) cross over and exchange segments, so a chromosome emerging from the process of meiosis may consist of some maternal (grandmaternal) and some paternal (grandpaternal) sections. If the probability of crossing over taking place is uniform along the length of a chromosome (which was later shown to be not quite true), then genes close together will be recombined less frequently than those far apart.

This realization opened an opportunity to map the arrangement of the genes and the estimated distances between them in the chromosome by studying the frequencies of recombination of various traits in the progenies of hybrids. In other words, the linkage maps of the chromosomes are really summaries of many statistical observations on the outcomes of hybridization experiments. In principle at least, such maps could be prepared even if the chromosomes, not to speak of the chiasmata at meiosis, were unknown. But an interesting and relevant fact is that in Drosophila males the linkage of the genes in the same chromosome is complete, and observations under the microscope show that no chiasmata are formed in the chromosomes at meiosis. In most organisms, including humans, chiasmata are seen in the meiotic chromosomes in both sexes, and observations on hybrid progenies show that recombination of linked genes occurs also in both sexes.

Chromosome maps exist for the Drosophila fly, corn, the house mouse, the bread mold Neurospora crassa, and some bacteria and bacteriophages (viruses that infect bacteria). Until quite late in the 20th century, the mapping of human chromosomes presented a particularly difficult problem: experimental crosses could not be arranged in humans, and only a few linkages could be determined by analysis of unique family histories. However, the development of recombinant DNA technology provided new understanding of human genetic processes and new methods of research. Using the techniques of recombinant DNA technology, hundreds of genes have been mapped to the human chromosomes and many linkages established.

Sex Linkage

The male of many animals has one chromosome pair, the sex chromosomes, consisting of unequal members called X and Y. At meiosis the X and Y chromosomes first pair then disjoin and pass to different cells. One-half of the gametes (spermatozoa) formed contain the X chromosome and the other half the Y. The female has two X chromosomes; all egg cells normally carry a single X. The eggs fertilized by X-bearing spermatozoa give females (XX), and those fertilized by Y-bearing spermatozoa give males (XY).

The genes located in the X chromosomes exhibit what is known as sex-linkage or crisscross inheritance. This is because of a crucial difference between the paired sex chromosomes and the other pairs of chromosomes (called autosomes). The members of the autosome pairs are truly homologous; that is, each member of a pair contains a full complement of the same genes (albeit, perhaps, in different allelic forms). The sex chromosomes, on the other hand, do not constitute a homologous pair, as the X chromosome is much larger and carries far more genes than does the Y. Consequently, many recessive alleles carried on the X chromosome of a male will be expressed just as if they were dominant, for the Y chromosome carries no genes to counteract them. The classic case of sex-linked inheritance, described by Morgan in 1910, is that of the white eyes in Drosophila. White-eyed females crossed to males with the normal red eye colour produce red-eyed daughters and white-eyed sons in the F1 generation and equal numbers of white-eyed and red-eyed females and males in the F2 generation.

The cross of red-eyed females to white-eyed males gives a different result: both sexes are red-eyed in F1 and the females in the F2 generation are red-eyed, half the males are red-eyed, and the other half white-eyed. As interpreted by Morgan, the gene that determines the red or white eyes is borne on the X chromosome, and the allele for red eye is dominant over that for white eye. Since a male receives its single X chromosome from his mother, all sons of white-eyed females also have white eyes. A female inherits one X chromosome from her mother and the other X from her father. Red-eyed females may have genes for red eyes in both of their X chromosomes (homozygotes), or they may have one X with the gene for red and the other for white (heterozygotes). In the progeny of heterozygous females, one-half of the sons will receive the X chromosome with the gene for white and will have white eyes, and the other half will receive the X with the gene for red eyes. The daughters of the heterozygous females crossed with white-eyed males will have either two X chromosomes with the gene for white—and hence have white eyes—or one X with the gene for white and the other X with the gene for red and will be red-eyed heterozygotes.

In humans, red-green colour blindness and hemophilia are among many traits showing sex-linked inheritance and are consequently due to genes borne in the X chromosome.

In some animals—birds, butterflies and moths, some fish, and at least some amphibians and reptiles—the chromosomal mechanism of sex determination is a mirror image of that described above. The male has two X chromosomes and the female an X and Y chromosome. Here the spermatozoa all have an X chromosome; the eggs are of two kinds, some with X and others with Y chromosomes, usually in equal numbers. The sex of the offspring is then determined by the egg rather than by the spermatozoon. Sex-linked inheritance is altered correspondingly. A male homozygous for a sex-linked recessive trait crossed to a female with the dominant one gives, in the F1 generation, daughters with the recessive trait and heterozygous sons with the corresponding dominant trait. The F2 generation has recessive and dominant females and males in equal numbers. A male with a dominant trait crossed to a female with a recessive trait gives uniformly dominant F1 and a segregation in a ratio of 2 dominant males: 1 dominant female: 1 recessive female.

Observations on pedigrees or experimental crosses show that certain traits exhibit sex-linked inheritance; the behaviour of the X chromosomes at meiosis is such that the genes they carry may be expected to exhibit sex-linkage. This evidence still failed to convince some skeptics that the genes for the sex-linked traits were in fact borne in certain chromosomes seen under the microscope. An experimental proof was furnished in 1916 by American geneticist Calvin Blackman Bridges. As stated above, white-eyed Drosophila females crossed to red-eyed males usually produce red-eyed female and white-eyed male progeny. Among thousands of such "regular" offspring, there are occasionally found exceptional white-eyed females and red-eyed males. Bridges constructed the following working hypothesis. Suppose that, during meiosis in the female, gametogenesis occasionally goes wrong, and the two X chromosomes fail to disjoin. Exceptional eggs will then be produced, carrying two X chromosomes and eggs carrying none. An egg with two X chromosomes coming from a white-eyed female fertilized by a spermatozoon with a Y chromosome will give an exceptional white-eyed female. An egg with no X chromosome fertilized by a spermatozoon with an X chromosome derived from a red-eyed father will yield an exceptional red-eyed male. This hypothesis can be rigorously tested. The exceptional white-eyed females should have not only the two X chromosomes but also a Y chromosome, which normal females do not have. The exceptional males should, on the other hand, lack a Y chromosome, which normal males do have. Both predictions were verified by examination under a microscope of the chromosomes of exceptional females and males. The hypothesis also predicts that exceptional eggs with two X chromosomes fertilized by X-bearing spermatozoa must give individuals with three X chromosomes; such individuals were later identified by Bridges as poorly viable "superfemales." Exceptional eggs with no Xs, fertilized by Y-bearing spermatozoa, will give zygotes without X chromosomes; such zygotes die in early stages of development.

Chromosomal Aberrations

The chromosome set of a species remains relatively stable over long periods of time. However, within populations there can be found abnormalities involving the structure or number of chromosomes. These alterations arise spontaneously from errors in the normal processes of the cell. Their consequences are usually deleterious, giving rise to individuals who are unhealthy or sterile, though in rare cases alterations provide new adaptive opportunities that allow evolutionary change to occur. In fact, the discovery of visible chromosomal differences between species has given rise to the belief that radical restructuring of chromosome architecture has been an important force in evolution.

Changes in Chromosome Structure

Two important principles dictate the properties of a large proportion of structural chromosomal changes. The first principle is that any deviation from the normal ratio of genetic material in the genome results in genetic imbalance and abnormal function. In the normal nuclei of both diploid and haploid cells, the ratio of the individual chromosomes to one another is 1:1. Any deviation from this ratio by addition or subtraction of either whole chromosomes or parts of chromosomes results in genomic imbalance. The second principle is that homologous chromosomes go to great lengths to pair at meiosis. The tightly paired homologous regions are joined by a ladderlike longitudinal structure called the synaptonemal complex. Homologous regions seem to be able to find each other and form a synaptonemal complex whether or not they are part of normal chromosomes. Therefore, when structural changes occur, not only are the resulting pairing formations highly characteristic of that type of structural change but they also dictate the packaging of normal and abnormal chromosomes into the gametes and subsequently into the progeny.

Deletions

The simplest, but perhaps most damaging, structural change is a deletion—the complete loss of a part of one chromosome. In a haploid cell this is lethal, because part of the essential genome is lost. However, even in diploid cells deletions are generally lethal or have other serious consequences. In a diploid a heterozygous deletion results in a cell that has one normal chromosome set and another set that contains a truncated chromosome. Such cells show genomic imbalance, which increases in severity with the size of the deletion. Another potential source of damage is that any recessive, deleterious, or lethal alleles that are in the normal counterpart of the deleted region will be expressed in the phenotype. In humans, cri-du-chat syndrome is caused by a heterozygous deletion at the tip of the short arm of chromosome 5. Infants are born with this condition as the result of a deletion arising in parental germinal tissues or even in sex cells. The manifestations of this deletion, in addition to the "cat cry" that gives the syndrome its name, include severe intellectual disability and an abnormally small head.

Duplications

A heterozygous duplication (an extra copy of some chromosome region) also results in a genomic imbalance with deleterious consequences. Small duplications within a gene can arise spontaneously. Larger duplications can be caused by crossovers following asymmetrical chromosome pairing or by meiotic irregularities resulting from other types of altered chromosome structures. If a duplication becomes homozygous, it can provide the organism with an opportunity to acquire new genetic functions through mutations within the duplicate copy.

Inversions

An inversion occurs when a chromosome breaks in two places and the region between the break rotates 180° before rejoining with the two end fragments. If the inverted segment contains the centromere (i.e., the point where the two chromatids are joined), the inversion is said to be pericentric; if not, it is called paracentric. Inversions do not result in a gain or loss of genetic material, and they have deleterious effects only if one of the chromosomal breaks occurs within an essential gene or if the function of a gene is altered by its relocation to a new chromosomal neighbourhood (called the position effect). However, individuals who are heterozygous for inversions produce aberrant meiotic products along with normal products. The only way uninverted and inverted segments can pair is by forming an inversion loop. If no crossovers occur in the loop, half of the gametes will be normal and the other half will contain an inverted chromosome. If a crossover does occur within the loop of a paracentric inversion, a chromosome bridge and an acentric chromosome (i.e., a chromosome without a centromere) will be formed, and this will give rise to abnormal meiotic products carrying deletions, which are inviable. In a pericentric inversion, a crossover within the loop does not result in a bridge or an acentric chromosome, but inviable products are produced carrying a duplication and a deletion.

Translocations

If a chromosome break occurs in each of two nonhomologous chromosomes and the two breaks rejoin in a new arrangement, the new segment is called a translocation. A cell bearing a heterozygous translocation has a full set of genes and will be viable unless one of the breaks causes damage within a gene or if there is a position effect on gene function. However, once again the pairing properties of the chromosomes at meiosis result in aberrant meiotic products. Specifically, half of the products are deleted for one of the chromosome regions that changed positions and half of the products are duplicated for the other. These duplications and deletions usually result in inviability, so translocation heterozygotes are generally semisterile ("half-sterile").

Changes in Chromosome Number

Two types of changes in chromosome numbers can be distinguished: a change in the number of whole chromosome sets (polyploidy) and a change in chromosomes within a set (aneuploidy).

Polyploids

An individual with additional chromosome sets is called a polyploid. Individuals with three sets of chromosomes (triploids, 3n) or four sets of chromosomes (tetraploids, 4n) are polyploid derivatives of the basic diploid (2n) constitution. Polyploids with odd numbers of sets (e.g., triploids) are sterile, because homologous chromosomes pair only two by two, and the extra chromosome moves randomly to a cell pole, resulting in highly unbalanced, nonfunctional meiotic products. It is for this reason that triploid watermelons are seedless. However, polyploids with even numbers of chromosome sets can be fertile if orderly two-by-two chromosome pairing occurs.

Though two organisms from closely related species frequently hybridize, the chromosomes of the fusing partners are different enough that the two sets do not pair at meiosis, resulting in sterile offspring. However, if by chance the number of chromosome sets in the hybrid accidentally duplicates, a pairing partner for each chromosome will be produced, and the hybrid will be fertile. These chromosomally doubled hybrids are called allotetraploids. Bread wheat, which is hexaploid (6n) due to several natural spontaneous hybridizations, is an example of an allotetraploid. Some polyploid plants are able to produce seeds through an asexual type of reproduction called apomixis; in such cases, all progeny are identical to the parent. Polyploidy does arise spontaneously in humans, but all polyploids either abort in utero or die shortly after birth.

Aneuploids

Some cells have an abnormal number of chromosomes that is not a whole multiple of the haploid number. This condition is called aneuploidy. Most aneuploids arise by nondisjunction, a failure of homologous chromosomes to separate at meiosis. When a gamete of this type is fertilized by a normal gamete, the zygotes formed will have an unequal distribution of chromosomes. Such genomic imbalance results in severe abnormalities or death. Only aneuploids involving small chromosomes tend to survive and even then only with an aberrant phenotype.

The most common form of aneuploidy in humans results in Down syndrome, a suite of specific disorders in individuals possessing an extra chromosome 21 (trisomy 21). The symptoms of Down syndrome include intellectual disability, severe disorders of internal organs such as the heart and kidneys, up-slanted eyes, an enlarged tongue, and abnormal dermal ridge patterns on the fingers, palms, and soles. Other forms of aneuploidy in humans result from abnormal numbers of sex chromosomes. Turner syndrome is a condition in which females have only one X chromosome. Symptoms may include short stature, webbed neck, kidney or heart malformations, underdeveloped sex characteristics, or sterility. Klinefelter syndrome is a condition in which males have one extra female sex chromosome, resulting in an XXY pattern. (Other, less frequent, chromosomal patterns include XXXY, XXXXY, XXYY, and XXXYY.) Symptoms of Klinefelter syndrome may include sterility, a tall physique, lack of secondary sex characteristics, breast development, and learning disabilities.

Variation

In Biology, Variation is any difference between cells, individual organisms, or groups of organisms of any species caused either by genetic differences (genotypic variation) or by the effect of environmental factors on the expression of the genetic potentials (phenotypic variation). Variation may be shown in physical appearance, metabolism, fertility, mode of reproduction, behaviour, learning and mental ability, and other obvious or measurable characters.

Genotypic variations are caused by differences in number or structure of chromosomes or by differences in the genes carried by the chromosomes. Eye colour, body form, and disease resistance are genotypic variations. Individuals with multiple sets of chromosomes are called polyploid; many common plants have two or more times the normal number of chromosomes, and new species may arise by this type of variation. A variation cannot be identified as genotypic by observation of the organism; breeding experiments must be performed under controlled environmental conditions to determine whether or not the alteration is inheritable.

Environmentally caused variations may result from one factor or the combined effects of several factors, such as climate, food supply, and actions of other organisms. Phenotypic variations also include stages in an organism's life cycle and seasonal variations in an individual. These variations do not involve any hereditary alteration and in general are not transmitted to future generations; consequently, they are not significant in the process of evolution.

Variations are classified either as continuous, or quantitative (smoothly grading between two extremes, with the majority of individuals at the centre, as height in human populations); or as discontinuous, or qualitative (composed of well-defined classes, as blood groups in man). A discontinuous variation with several classes, none of which is very small, is known as a polymorphic variation. The separation of most higher organisms into males and females and the occurrence of several forms of a butterfly of the same species, each coloured to blend with a different vegetation, are examples of polymorphic variation.

Genetic Variation

Genetic variation is a measure of the genetic differences that exist within a population. The genetic variation of an entire species is often called genetic diversity. Genetic variations are the differences in DNA segments or genes between individuals and each variation of a gene is called an allele. For example, a population with many different alleles at a single chromosome locus has a high amount of genetic variation. Genetic variation is essential for natural selection because natural selection can only increase or decrease frequency of alleles that already exist in the population.

Genetic variation is caused by:

- Mutation,
- Random mating between organisms,
- Random fertilization,
- Crossing over (or recombination) between chromatids of homologous chromosomes during meiosis.

The last three of these factors reshuffle alleles within a population, giving offspring combinations which differ from their parents and from others.

Genetic variation in the shells of Donax variabilis: An enormous amount of phenotypic variation exists in the shells of Donax varabilis, otherwise known as the coquina mollusc. This phenotypic variation is due at least partly to genetic variation within the coquina population.

Evolution and Adaptation to the Environment

Variation allows some individuals within a population to adapt to the changing environment. Because natural selection acts directly only on phenotypes, more genetic variation within a population usually enables more phenotypic variation. Some new alleles increase an organism's ability to survive and reproduce, which then ensures the survival of the allele in the population. Other new alleles may be immediately detrimental (such as a malformed oxygen-carrying protein) and organisms carrying these new mutations will die out. Neutral alleles are neither selected for nor against and usually remain in the population. Genetic variation is advantageous because it enables some individuals and, therefore, a population, to survive despite a changing environment.

Low genetic diversity in the wild cheetah population: Populations of wild cheetahs have very low genetic variation. Because wild cheetahs are threatened, their species has a very low genetic diversity. This low genetic diversity means they are often susceptible to disease and often pass on lethal recessive mutations; only about 5% of cheetahs survive to adulthood.

Geographic Variation

Some species display geographic variation as well as variation within a population. Geographic variation, or the distinctions in the genetic makeup of different populations, often occurs when populations are geographically separated by environmental barriers or when they are under selection pressures from a different environment. One example of geographic variation are clines: graded changes in a character down a geographic axis.

Sources of Genetic Variation

Gene duplication, mutation, or other processes can produce new genes and alleles and increase genetic variation. New genetic variation can be created within generations in a population, so a population with rapid reproduction rates will probably have high genetic variation. However, existing genes can be arranged in new ways from chromosomal crossing over and recombination in sexual reproduction. Overall, the main sources of genetic variation are the formation of new alleles, the altering of gene number or position, rapid reproduction, and sexual reproduction.

Mutation

Mutation is an alteration in the genetic material (the genome) of a cell of a living organism or of a virus that is more or less permanent and that can be transmitted to the cell's or the virus's descendants. (The genomes of organisms are all composed of DNA, whereas viral genomes can be of DNA or RNA.) Mutation in the DNA of a body cell of a multicellular organism (somatic mutation) may be transmitted to descendant cells by DNA replication and hence result in a sector or patch of cells having abnormal function, an example being cancer. Mutations in egg or sperm cells (germinal mutations) may result in an individual offspring all of whose cells carry the mutation, which often confers some serious malfunction, as in the case of a human genetic disease such as cystic fibrosis. Mutations result either from accidents during the normal chemical transactions of DNA, often during replication, or from exposure to high-energy electromagnetic radiation (e.g., ultraviolet light or X-rays) or particle radiation or to highly reactive chemicals in the environment. Because mutations are random changes, they are expected to be mostly deleterious, but some may be beneficial in certain environments. In general, mutation is the main source of genetic variation, which is the raw material for evolution by natural selection.

The genome is composed of one to several long molecules of DNA, and mutation can occur potentially anywhere on these molecules at any time. The most serious changes take place in the functional units of DNA, the genes. A mutated form of a gene is called a mutant allele. A gene is typically composed of a regulatory region, which is responsible for turning the gene's transcription on and off at the appropriate times during development, and a coding region, which carries the genetic code for the structure of a functional molecule, generally a protein. A protein is a chain of usually several hundred amino acids. Cells make 20 common amino acids, and it is the unique number and sequence of these that give a protein its specific function. Each amino acid is encoded by a unique sequence, or codon, of three of the four possible base pairs in the DNA (A–T, T–A, G–C, and C–G, the individual letters referring to the four nitrogenous bases adenine, thymine, guanine, and cytosine). Hence, a mutation that changes DNA sequence can change amino acid sequence and in this way potentially reduce or inactivate a protein's function. A change in the DNA sequence of a gene's regulatory region can adversely affect the timing and availability of the gene's protein and also lead to serious cellular malfunction. On the other hand, many mutations are silent, showing no obvious effect at the functional level. Some silent mutations are in the DNA between genes, or they are of a type that results in no significant amino acid changes.

Mutations are of several types. Changes within genes are called point mutations. The simplest kinds are changes to single base pairs, called base-pair substitutions. Many of these substitute an incorrect amino acid in the corresponding position in the encoded protein, and of these a large proportion result in altered protein function. Some base-pair substitutions produce a stop codon. Normally, when a stop codon occurs at the end of a gene, it stops protein synthesis, but, when it occurs in an abnormal position, it can result in a truncated and nonfunctional protein. Another type of simple change, the deletion or insertion of single base pairs, generally has a profound effect on the protein because the protein's synthesis, which is carried out by the reading of triplet codons in a linear fashion from one end of the gene to the other, is thrown off. This change leads to a frameshift in reading the gene such that all amino acids are incorrect from the mutation onward. More-complex combinations of base substitutions, insertions, and deletions can also be observed in some mutant genes.

Mutations that span more than one gene are called chromosomal mutations because they affect the structure, function, and inheritance of whole DNA molecules (microscopically visible in a coiled state as chromosomes). Often these chromosome mutations result from one or more coincident breaks in the DNA molecules of the genome (possibly from exposure to energetic radiation), followed in some cases by faulty rejoining. Some outcomes are large-scale deletions, duplications, inversions, and translocations. In a diploid species (a species, such as human beings, that has a double set of chromosomes in the nucleus of each cell), deletions and duplications alter gene balance and often result in abnormality. Inversions and translocations involve no loss or gain and are functionally normal unless a break occurs within a gene. However, at meiosis (the specialized nuclear divisions that take place during the production of gametes—i.e., eggs and sperm), faulty pairing of an inverted or translocated chromosome set with a normal set can result in gametes and hence progeny with duplications and deletions.

Loss or gain of whole chromosomes results in a condition called aneuploidy. One familiar result of aneuploidy is Down syndrome, a chromosomal disorder in which humans are born with an extra chromosome 21 (and hence bear three copies of that chromosome instead of the usual two). Another type of chromosome mutation is the gain or loss of whole chromosome sets. Gain of sets results in polyploidy—that is, the presence of three, four, or more chromosome sets instead of the usual two. Polyploidy has been a significant force in the evolution of new species of plants and animals.

Most genomes contain mobile DNA elements that move from one location to another. The movement of these elements can cause mutation, either because the element arrives in some crucial location, such as within a gene, or because it promotes large-scale chromosome mutations via recombination between pairs of mobile elements in different locations.

At the level of whole populations of organisms, mutation can be viewed as a constantly dripping faucet introducing mutant alleles into the population, a concept described as mutational pressure. The rate of mutation differs for different genes and organisms. In RNA viruses, such as the human immunodeficiency virus, replication of the genome takes place within the host cell using a mechanism that is prone to error. Hence, mutation rates in such viruses are high. In general, however, the fate of individual mutant alleles is never certain. Most are eliminated by chance. In some cases a mutant allele can increase in frequency by chance, and then individuals expressing the allele can be subject to selection, either positive or negative. Hence, for any one gene the frequency of a mutant allele in a population is determined by a combination of mutational pressure, selection, and chance.

References

- Biological-inheritance-meaning-categories-specifications, mendels-laws, biochemistry: biologydiscussion.com, Retrieved 2 April, 2019
- Mendel-and-his-peas, introduction-to-heredity, classical-genetics, science: khanacademy.org, Retrieved 12 July, 2019
- Laws-of-inheritance, biology: lumenlearning.com, Retrieved 6 January, 2019
- Heredity, biology: toppr.com, Retrieved 16 July, 2019
- Heredity-genetics, science: britannica.com, Retrieved 13 May, 2019
- Variation-biology, science: britannica.com, Retrieved 19 February, 2019
- Genetic-Variation, Population-Genetics, Evolution-of-Populations, Introductory-and-General-Biology: libretexts.org, Retrieved 1 June, 2019
- Mutation-genetics, science: britannica.com, Retrieved 23 August, 2019

Chapter 5
Theories and Concepts of Evolution

The change in genetic characteristics of biological population over consecutive generations is known as evolution. These characteristics are the expressions of genes that are passed on from parent to offspring during reproduction. Theories of evolution include common descent, coevolution, speciation, adaptation, etc. This chapter discusses in detail these theories and methodologies related to evolution.

Evolution

Evolution is the theory in biology postulating that the various types of plants, animals, and other living things on Earth have their origin in other preexisting types and that the distinguishable differences are due to modifications in successive generations. The theory of evolution is one of the fundamental keystones of modern biological theory.

The diversity of the living world is staggering. More than 2 million existing species of organisms have been named and described; many more remain to be discovered—from 10 million to 30 million, according to some estimates. What is impressive is not just the numbers but also the incredible heterogeneity in size, shape, and way of life—from lowly bacteria, measuring less than a thousandth of a millimetre in diameter, to stately sequoias, rising 100 metres (300 feet) above the ground and weighing several thousand tons; from bacteria living in hot springs at temperatures near the boiling point of water to fungi and algae thriving on the ice masses of Antarctica and in saline pools at −23 °C (−9 °F); and from giant tube worms discovered living near hydrothermal vents on the dark ocean floor to spiders and larkspur plants existing on the slopes of Mount Everest more than 6,000 metres (19,700 feet) above sea level.

The virtually infinite variations on life are the fruit of the evolutionary process. All living creatures are related by descent from common ancestors. Humans and other mammals descend from shrewlike creatures that lived more than 150 million years ago; mammals, birds, reptiles, amphibians, and fishes share as ancestors aquatic worms that lived 600 million years ago; and all plants and animals derive from bacteria-like microorganisms that originated more than 3 billion years ago. Biological evolution is a process of descent with modification. Lineages of organisms change through generations; diversity arises because the lineages that descend from common ancestors diverge through time.

The 19th-century English naturalist Charles Darwin argued that organisms come about by evolution, and he provided a scientific explanation, essentially correct but incomplete, of how evolution occurs and why it is that organisms have features—such as wings, eyes, and kidneys—clearly structured to serve specific functions. Natural selection was the fundamental concept in his explanation. Natural selection occurs because individuals having more-useful traits, such as more-acute vision or swifter legs, survive better and produce more progeny than individuals with less-favourable traits. Genetics, a science born in the 20th century, reveals in detail how natural selection works and led to the development of the modern theory of evolution. Beginning in the 1960s, a related scientific discipline, molecular biology, enormously advanced knowledge of biological evolution and made it possible to investigate detailed problems that had seemed completely out of reach only a short time previously—for example, how similar the genes of humans and chimpanzees might be (they differ in about 1–2 percent of the units that make up the genes).

The Evidence for Evolution

Darwin and other 19th-century biologists found compelling evidence for biological evolution in the comparative study of living organisms, in their geographic distribution, and in the fossil remains of extinct organisms. Since Darwin's time, the evidence from these sources has become considerably stronger and more comprehensive, while biological disciplines that emerged more recently—genetics, biochemistry, physiology, ecology, animal behaviour (ethology), and especially molecular biology—have supplied powerful additional evidence and detailed confirmation. The amount of information about evolutionary history stored in the DNA and proteins of living things is virtually unlimited; scientists can reconstruct any detail of the evolutionary history of life by investing sufficient time and laboratory resources.

Evolutionists no longer are concerned with obtaining evidence to support the fact of evolution but rather are concerned with what sorts of knowledge can be obtained from different sources of evidence. The following sections identify the most productive of these sources and illustrate the types of information they have provided.

The Fossil Record

Paleontologists have recovered and studied the fossil remains of many thousands of organisms that lived in the past. This fossil record shows that many kinds of extinct organisms were very different in form from any now living. It also shows successions of organisms through time, manifesting their transition from one form to another.

When an organism dies, it is usually destroyed by other forms of life and by weathering processes. On rare occasions some body parts—particularly hard ones such as shells, teeth, or bones—are preserved by being buried in mud or protected in some other way from predators and weather. Eventually, they may become petrified and preserved indefinitely with the rocks in which they are embedded. Methods such as radiometric dating—measuring the amounts of natural radioactive atoms that remain in certain minerals to determine the elapsed time since they were constituted—make it possible to estimate the time period when the rocks, and the fossils associated with them, were formed.

Radiometric dating indicates that Earth was formed about 4.5 billion years ago. The earliest fossils resemble microorganisms such as bacteria and cyanobacteria (blue-green algae); the oldest of these fossils appear in rocks 3.5 billion years old. The oldest known animal fossils, about 700 million years old, come from the so-called Ediacara fauna, small wormlike creatures with soft bodies. Numerous fossils belonging to many living phyla and exhibiting mineralized skeletons appear in rocks about 540 million years old. These organisms are different from organisms living now and from those living at intervening times. Some are so radically different that paleontologists have created new phyla in order to classify them. The first vertebrates, animals with backbones, appeared about 400 million years ago; the first mammals, less than 200 million years ago. The history of life recorded by fossils presents compelling evidence of evolution.

The fossil record is incomplete. Of the small proportion of organisms preserved as fossils, only a tiny fraction have been recovered and studied by paleontologists. In some cases the succession of forms over time has been reconstructed in detail. One example is the evolution of the horse. The horse can be traced to an animal the size of a dog having several toes on each foot and teeth appropriate for browsing; this animal, called the dawn horse (genus Hyracotherium), lived more than 50 million years ago. The most recent form, the modern horse (Equus), is much larger in size, is one-toed, and has teeth appropriate for grazing. The transitional forms are well preserved as fossils, as are many other kinds of extinct horses that evolved in different directions and left no living descendants.

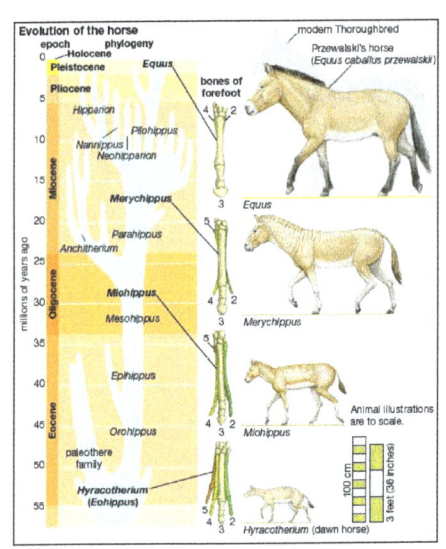

Evolution of the horse.

Evolution of the horse over the past 55 million years. The present-day Przewalski's horse is believed to be the only remaining example of a wild horse—i.e., the last remaining modern horse to have evolved by natural selection. Numbered bones in the forefoot illustrations trace the gradual transition from a four-toed to a one-toed animal.

Using recovered fossils, paleontologists have reconstructed examples of radical evolutionary transitions in form and function. For example, the lower jaw of reptiles contains several bones, but that of mammals only one. The other bones in the reptile jaw unmistakably evolved into bones now found in the mammalian ear. At first, such a transition would seem unlikely—it is hard to imagine what function such bones could have had during their intermediate stages. Yet paleontologists discovered two transitional forms of mammal-like reptiles, called therapsids, that had a double jaw joint (i.e., two hinge points side by side)—one joint consisting of the bones that persist in the mammalian jaw and the other composed of the quadrate and articular bones, which eventually became the hammer and anvil of the mammalian ear.

For skeptical contemporaries of Darwin, the "missing link"—the absence of any known transitional form between apes and humans—was a battle cry, as it remained for uninformed people afterward.

Not one but many creatures intermediate between living apes and humans have since been found as fossils. The oldest known fossil hominins—i.e., primates belonging to the human lineage after it separated from lineages going to the apes—are 6 million to 7 million years old, come from Africa, and are known as Sahelanthropus and Orrorin (or Praeanthropus), which were predominantly bipedal when on the ground but which had very small brains. Ardipithecus lived about 4.4 million years ago, also in Africa. Numerous fossil remains from diverse African origins are known of Australopithecus, a hominin that appeared between 3 million and 4 million years ago.

Australopithecus had an upright human stance but a cranial capacity of less than 500 cc (equivalent to a brain weight of about 500 grams), comparable to that of a gorilla or a chimpanzee and about one-third that of humans. Its head displayed a mixture of ape and human characteristics—a low forehead and a long, apelike face but with teeth proportioned like those of humans. Other early hominins partly contemporaneous with Australopithecus include Kenyanthropus and Paranthropus; both had comparatively small brains, although some species of Paranthropus had larger bodies. Paranthropus represents a side branch in the hominin lineage that became extinct. Along with increased cranial capacity, other human characteristics have been found in Homo habilis, which lived about 1.5 million to 2 million years ago in Africa and had a cranial capacity of more than 600 cc (brain weight of 600 grams), and in H. erectus, which lived between 0.5 million and more than 1.5 million years ago, apparently ranged widely over Africa, Asia, and Europe, and had a cranial capacity of 800 to 1,100 cc (brain weight of 800 to 1,100 grams). The brain sizes of H. ergaster, H. antecessor, and H. heidelbergensis were roughly that of the brain of H. erectus, some of which species were partly contemporaneous, though they lived in different regions of the Eastern Hemisphere.

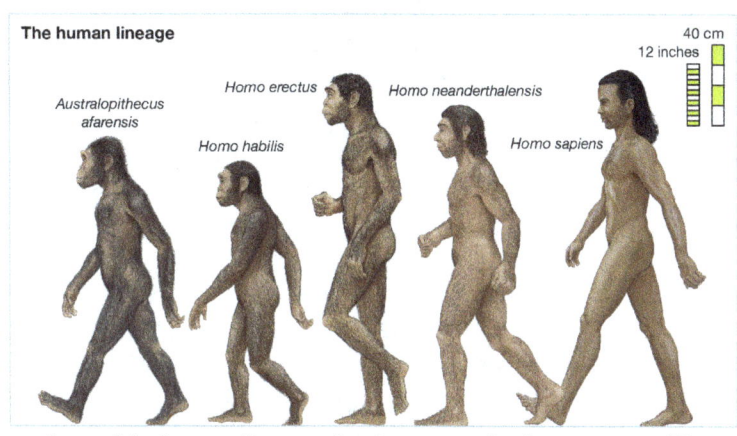

Five hominins—members of the human lineage after it separated at least seven million to six million years ago from lineages going to the apes—are depicted in an artist's interpretations. All but Homo sapiens, the species that comprises modern humans, are extinct and have been reconstructed from fossil evidence.

Structural Similarities

The skeletons of turtles, horses, humans, birds, and bats are strikingly similar, in spite of the different ways of life of these animals and the diversity of their environments. The correspondence, bone by bone, can easily be seen not only in the limbs but also in every other part of the body. From a purely practical point of view, it is incomprehensible that a turtle should swim, a horse run, a person write, and a bird or a bat fly with forelimb structures built of the same bones. An engineer could design better limbs in each case. But if it is accepted that all of these skeletons inherited their structures from a common ancestor and became modified only as they adapted to different ways of life, the similarity of their structures makes sense.

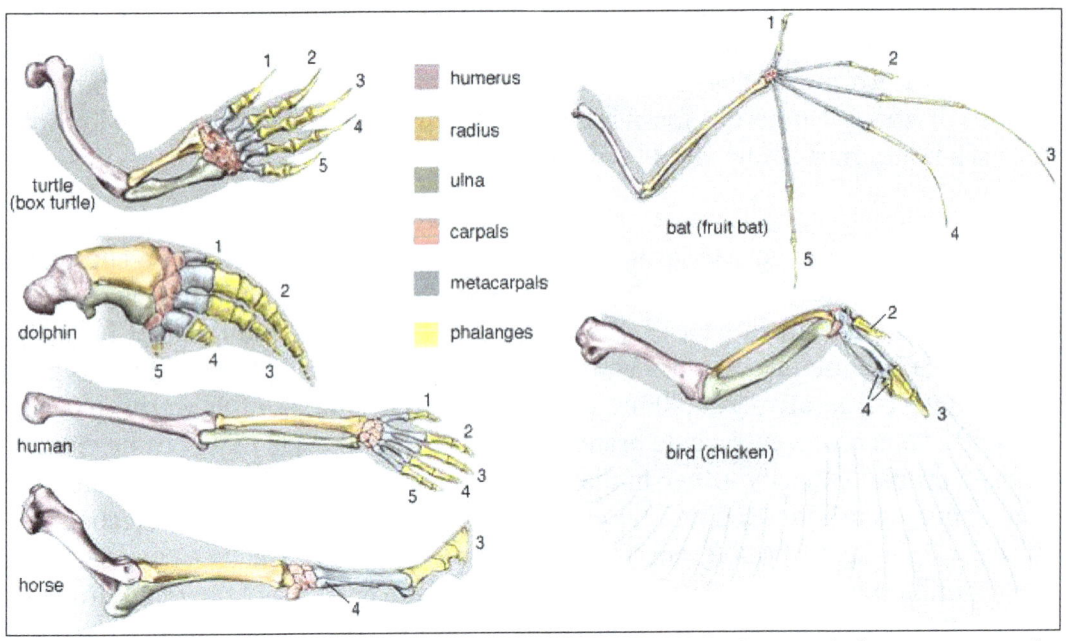

Homologies of the forelimb among vertebrates, giving evidence for evolution. The bones correspond, although they are adapted to the specific mode of life of the animal. (Some anatomists interpret the digits in the bird's wing as being 1, 2, and 3, rather than 2, 3, and 4.)

Comparative anatomy investigates the homologies, or inherited similarities, among organisms in bone structure and in other parts of the body. The correspondence of structures is typically very close among some organisms—the different varieties of songbirds, for instance—but becomes less so as the organisms being compared are less closely related in their evolutionary history. The similarities are less between mammals and birds than they are among mammals, and they are still less between mammals and fishes. Similarities in structure, therefore, not only manifest evolution but also help to reconstruct the phylogeny, or evolutionary history, of organisms.

Comparative anatomy also reveals why most organismic structures are not perfect. Like the forelimbs of turtles, horses, humans, birds, and bats, an organism's body parts are less than perfectly adapted because they are modified from an inherited structure rather than designed from completely "raw" materials for a specific purpose. The imperfection of structures is evidence for evolution and against antievolutionist arguments that invoke intelligent design.

Embryonic Development and Vestiges

Darwin and his followers found support for evolution in the study of embryology, the science that investigates the development of organisms from fertilized egg to time of birth or hatching. Vertebrates, from fishes through lizards to humans, develop in ways that are remarkably similar during early stages, but they become more and more differentiated as the embryos approach maturity. The similarities persist longer between organisms that are more closely related (e.g., humans and monkeys) than between those less closely related (humans and sharks). Common developmental patterns reflect evolutionary kinship. Lizards and humans share a developmental pattern inherited from their remote common ancestor; the inherited pattern of each was modified only as the separate descendant lineages evolved in different directions. The common embryonic stages of the two creatures reflect the constraints imposed by this common inheritance, which prevents changes that have not been necessitated by their diverging environments and ways of life.

The embryos of humans and other nonaquatic vertebrates exhibit gill slits even though they never breathe through gills. These slits are found in the embryos of all vertebrates because they share as common ancestors the fish in which these structures first evolved. Human embryos also exhibit by the fourth week of development a well-defined tail, which reaches maximum length at six weeks. Similar embryonic tails are found in other mammals, such as dogs, horses, and monkeys; in humans, however, the tail eventually shortens, persisting only as a rudiment in the adult coccyx.

A close evolutionary relationship between organisms that appear drastically different as adults can sometimes be recognized by their embryonic homologies. Barnacles, for example, are sedentary crustaceans with little apparent likeness to such free-swimming crustaceans as lobsters, shrimps, or copepods. Yet barnacles pass through a free-swimming larval stage, the nauplius, which is unmistakably similar to that of other crustacean larvae.

Embryonic rudiments that never fully develop, such as the gill slits in humans, are common in all sorts of animals. Some, however, like the tail rudiment in humans, persist as adult vestiges, reflecting evolutionary ancestry. The most familiar rudimentary organ in humans is the vermiform appendix. This wormlike structure attaches to a short section of intestine called the cecum, which is located at the point where the large and small intestines join. The human vermiform appendix is a functionless vestige of a fully developed organ present in other mammals, such as the rabbit and other herbivores, where a large cecum and appendix store vegetable cellulose to enable its digestion with the help of bacteria. Vestiges are instances of imperfections—like the imperfections seen in anatomical structures—that argue against creation by design but are fully understandable as a result of evolution.

Biogeography

Darwin also saw a confirmation of evolution in the geographic distribution of plants and animals, and later knowledge has reinforced his observations. For example, there are about 1,500 known species of Drosophila vinegar flies in the world; nearly one-third of them live in Hawaii and nowhere else, although the total area of the archipelago is less than one-twentieth the area of California or Germany. Also in Hawaii are more than 1,000 species of snails and other land mollusks that exist nowhere else. This unusual diversity is easily explained by evolution. The islands of Hawaii are extremely isolated and have had few colonizers—i.e, animals and plants that arrived there from elsewhere and established populations. Those species that did colonize the islands found many unoccupied ecological niches, local environments suited to sustaining them and lacking predators that would prevent them from multiplying. In response, these species rapidly diversified; this process of diversifying in order to fill ecological niches is called adaptive radiation.

Each of the world's continents has its own distinctive collection of animals and plants. In Africa are rhinoceroses, hippopotamuses, lions, hyenas, giraffes, zebras, lemurs, monkeys with narrow noses and nonprehensile tails, chimpanzees, and gorillas. South America, which extends over much the same latitudes as Africa, has none of these animals; it instead has pumas, jaguars, tapir, llamas, raccoons, opossums, armadillos, and monkeys with broad noses and large prehensile tails.

These vagaries of biogeography are not due solely to the suitability of the different environments. There is no reason to believe that South American animals are not well suited to living in Africa or those of Africa to living in South America. The islands of Hawaii are no better suited than other Pacific islands for vinegar flies, nor are they less hospitable than other parts of the world for many absent organisms. In fact, although no large mammals are native to the Hawaiian islands, pigs and goats have multiplied there as wild animals since being introduced by humans. This absence of many species from a hospitable environment in which an extraordinary variety of other species flourish can be explained by the theory of evolution, which holds that species can exist and evolve only in geographic areas that were colonized by their ancestors.

Molecular Biology

The field of molecular biology provides the most detailed and convincing evidence available for biological evolution. In its unveiling of the nature of DNA and the workings of organisms at the level of enzymes and other protein molecules, it has shown that these molecules hold information about an organism's ancestry. This has made it possible to reconstruct evolutionary events that were previously unknown and to confirm and adjust the view of events already known. The precision with which these events can be reconstructed is one reason the evidence from molecular biology is so compelling. Another reason is that molecular evolution has shown all living organisms, from bacteria to humans, to be related by descent from common ancestors.

A remarkable uniformity exists in the molecular components of organisms—in the nature of the components as well as in the ways in which they are assembled and used. In all bacteria, plants, animals, and humans, the DNA comprises a different sequence of the same four component nucleotides, and all the various proteins are synthesized from different combinations and sequences of the same 20 amino acids, although several hundred other amino acids do exist. The genetic code by which the information contained in the DNA of the cell nucleus is passed on to proteins is virtually everywhere the same. Similar metabolic pathways—sequences of biochemical reactions—are used by the most diverse organisms to produce energy and to make up the cell components.

This unity reveals the genetic continuity and common ancestry of all organisms. There is no other rational way to account for their molecular uniformity when numerous alternative structures are equally likely. The genetic code serves as an example. Each particular sequence of three nucleotides in the nuclear DNA acts as a pattern for the production of exactly the same amino acid in all organisms. This is no more necessary than it is for a language to use a particular combination of letters to represent a particular object. If it is found that certain sequences of letters—planet, tree, woman—are used with identical meanings in a number of different books, one can be sure that the languages used in those books are of common origin.

Genes and proteins are long molecules that contain information in the sequence of their components in much the same way as sentences of the English language contain information in the sequence of their letters and words. The sequences that make up the genes are passed on from parents to offspring and are identical except for occasional changes introduced by mutations. As an illustration, one may assume that two books are being compared. Both books are 200 pages long and contain the same number of chapters. Closer examination reveals that the two books are identical page for page and word for word, except that an occasional word—say, one in 100—is different. The two books cannot have been written independently; either one has been copied from the other, or both have been copied, directly or indirectly, from the same original book. Similarly, if each component nucleotide of DNA is represented by one letter, the complete sequence of nucleotides in the DNA of a higher organism would require several hundred books of hundreds of pages, with several thousand letters on each page. When the "pages" (or sequences of nucleotides) in these "books" (organisms) are examined one by one, the correspondence in the "letters" (nucleotides) gives unmistakable evidence of common origin.

The two arguments presented above are based on different grounds, although both attest to evolution. Using the alphabet analogy, the first argument says that languages that use the same dictionary—the same genetic code and the same 20 amino acids—cannot be of independent origin. The second argument, concerning similarity in the sequence of nucleotides in the DNA (and thus the sequence of amino acids in the proteins), says that books with very similar texts cannot be of independent origin.

The evidence of evolution revealed by molecular biology goes even farther. The degree of similarity in the sequence of nucleotides or of amino acids can be precisely quantified. For example, in humans and chimpanzees, the protein molecule called cytochrome c, which serves a vital function in respiration within cells, consists of the same 104 amino acids in exactly the same order. It differs, however, from the cytochrome c of rhesus monkeys by 1 amino acid, from that of horses by 11 additional amino acids, and from that of tuna by 21 additional amino acids. The degree of similarity reflects the recency of common ancestry. Thus, the inferences from comparative anatomy and other disciplines concerning evolutionary history can be tested in molecular studies of DNA and proteins by examining their sequences of nucleotides and amino acids.

The authority of this kind of test is overwhelming; each of the thousands of genes and thousands of proteins contained in an organism provides an independent test of that organism's evolutionary history. Not all possible tests have been performed, but many hundreds have been done, and not one has given evidence contrary to evolution. There is probably no other notion in any field of science that has been as extensively tested and as thoroughly corroborated as the evolutionary origin of living organisms.

Convergent Evolution

Convergent evolution is the process by which unrelated or distantly related organisms evolve similar body forms, coloration, organs, and adaptations. Natural selection can result in evolutionary convergence under several different circumstances. Species can converge in sympatry, as in mimicry complexes among insects, especially butterflies (coral snakes and their mimics constitute another well-known example). Mimicry evolves after one species, the 'model' has become aposematic (warningly colored) because it is toxic or poisonous and therefore protected. Two distinct kinds of mimicry are recognized, Batesian and Müllerian. In Batesian mimicry, the mimic is palatable or unprotected, but gains from being mistaken for the model, which is unpalatable or protected. Two protected model species can also converge because of the advantage of being mistaken for each other (Müllerian mimicry).

Viceroy butterfly (the Mimic - palatable species)
Monarch butterfly (The model - distasteful species)

Mimicry is an interesting consequence of warning coloration that nicely demonstrates the power of natural selection. An organism that commonly occurs in a community along with a poisonous or distasteful species can benefit from a resemblance to the warningly colored species, even though the 'mimic' itself is nonpoisonous and/or quite palatable. Because predators that have experienced contacts with the model species, and have learned to avoid it, mistake the mimic species for the model and avoid it as well. Such false warning coloration is termed Batesian mimicry after its discoverer.

Eastern Coral Snake (venomous)
Scarlet King Snake (non-venomous)

Many species of harmless snakes mimic poisonous snakes; in Central America, some harmless snakes are so similar to poisonous coral snakes that only an expert can distinguish the mimic from the 'model.' A few experts have even died as a result of a superficial misidentifications. Similarly, certain harmless flies and clearwing moths mimic bees and wasps, and palatable species of butterflies mimic distasteful species. Batesian mimicry is disadvantageous to the model species because some predators will encounter palatable or harmless mimics and thereby take longer to learn to avoid the model. The greater the proportion of mimics to models, the longer is the time required for predator learning and the greater the number of model casualties. In fact, if mimics became more abundant than models, predators might not learn to avoid the prey item at all but might actively search out model and mimic alike. For this reason Batesian mimics are usually much less abundant than their models; also, mimics of this sort are frequently polymorphic (often only females are mimics) and mimic several different model species.

Müllerian mimicry is different, and occurs when two species, both distasteful or dangerous, mimic one another. Both bees and wasps, for example, are usually banded with yellows and blacks. Because potential predators encounter several species of Müllerian mimics more frequently than just a single species, they learn to avoid them faster, and the relationship is actually beneficial to both prey species. The resemblance need not be as precise as it must be under Batesian mimicry because neither species actually deceives the predator; rather, each only reminds the predator of its dangerous or distasteful properties. Müllerian mimicry is beneficial to all parties including the predator; mimics can be equally common and are rarely polymorphic.

Molecules can evolve convergently, especially when parasites mimic molecular messages that signal 'self' to immune responses of hosts, which allows the parasite to elude its host's defenses. Molecular convergence could also take place when a particular metabolic function requires similar or identical molecular structure. Some gene circuits and gene networks appear to have undergone convergent evolution by single-gene duplications in higher eukaryotes. Convergence in DNA nucleotide sequences would lead to erroneous phylogenetic conclusions, which would be problematical for molecular systematic studies.

Evolutionary convergence involving unrelated organisms living in similar environments but in different places (allopatry) can also occur in another way. This usually takes place in relatively simple communities in which biotic interactions are highly predictable and the resulting number of different ways of exploiting the environment are limited. Similar environments pose similar challenges to survival and reproduction, and those traits that enhance Darwinian fitness are selected for in each environment. Such organisms that fill similar ecological roles in different, independently-evolved, biotas are termed "ecological equivalents". Examples are legion.

Wings and wing-like structures have evolved independently several times, in insects, reptiles (pterosaurs and birds) and in mammals (bats). Flight first evolved in insects about 330 million years ago (mya), second in pterosaurs (about 225 mya), later in birds (about 150 mya), and still later in bats (50-60 mya). Some frogs, lizards, and mammals have also evolved the ability to glide, presumably a precursor to flight. In order to land safely, such hang gliders must time their stall precisely at the right moment and place.

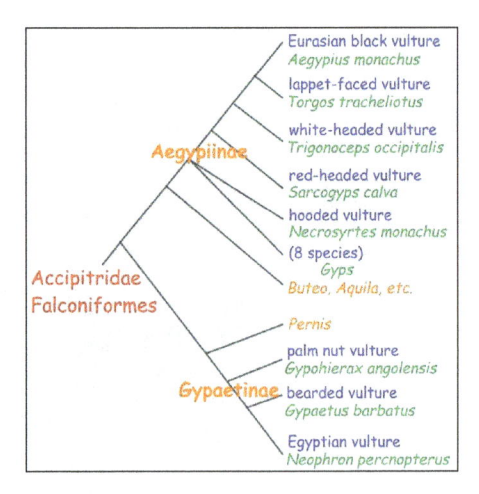

```
                                    Eurasian black vulture
                                     Aegypius monachus
                                    lappet-faced vulture
                                     Torgos tracheliotus
                                    white-headed vulture
                        Aegypiinae  Trigonoceps occipitalis
                                    red-headed vulture
                                     Sarcogyps calva
                                    hooded vulture
                                     Necrosyrtes monachus
                                    (8 species)
         Accipitridae                Gyps
         Falconiformes              Buteo, Aquila, etc.

                                    Pernis
                                    palm nut vulture
                                     Gypohierax angolensis
                        Gypaetinae  bearded vulture
                                     Gypaetus barbatus
                                    Egyptian vulture
                                     Neophron percnopterus
```

For many years, avian systematists classified Old World and New World vultures as close relatives, both thought to be allied to raptors (hawks and owls). However, DNA hybridization suggested that, although Old World vultures are indeed related to raptors, New World vultures are not, but are descendents of common ancestors to storks and cranes (more recent studies are equivocal but still support independent evolution of the two clades). Morphological convergence was strong enough to actually mislead students of bird classification. Interestingly, a behavioral trait was conserved in the evolution of new world vultures: when heat stressed, storks defecate/urinate on their own legs to dissipate excess heat. New World vultures do this, whereas Old World vultures do not.

New World Cactus African Euphorb

Arid regions of South Africa support a wide variety of euphorbeaceous plants, some of which are strikingly close to American cacti phenotypically. They are leafless stem succulents, protected by sharp spines, presumably adaptations to reduce water loss and predation in arid environments. Similarly, evergreen sclerophyll woody shrubs have evolved convergently under Mediterranean climates in several different regions.

African Macronix American Medolark

A brown bird of some African prairies and grasslands, the African yellow-throated longclaw (Macronix croceus), a motacillid, has a yellow breast with a black chevron "V". This motacillid looks and acts so much like an American meadowlark (Sturnella magna), an icterid, that a competent bird watcher might mistake them for the same species, yet they belong to different avian families. Another example is the North American Little Auk and the Magellan Diving Petrel, two superficially very similar aquatic birds, which belong to different avian orders.

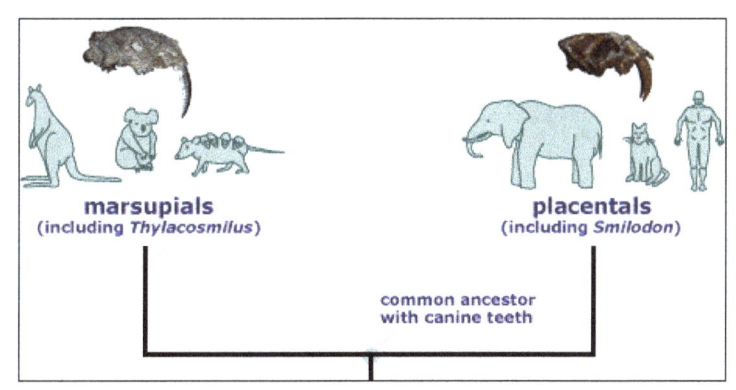

South American Thylacosmilus   North American Smilodon

Convergence sometimes occurs under unusual conditions where selective forces for the achievement of a particular mode of existence are particularly strong. Presumably in response to thick-skinned prey, two fossil saber-tooth carnivores, the South American marsupial 'cat,' Thylacosmilus, and the North American placental saber-toothed tiger, Smilodon, evolved long knife-like canine teeth independently (but these were not contemporary). Many other marsupial mammals have undergone convergent evolution with placentals, including moles, mice, wombats (woodchucks), numbats (anteaters), quolls (cats), and thylacines (wolves).

Still another example of convergent evolution is seen in the similar shape and coloration of fish and cetaceans, both of which have adapted to the marine environment by developing a fusiform body and neutral buoyancy (an extinct group of marine reptiles known as ichthyosaurs evolved the same body plan). Sharks and dolphins are also countershaded, with a light underbelly and a darker upper surface, which makes them less visible from both below and above. However, countershading is actually the rule among both arthropods and vertebrates, so it is presumably an ancestral state that has been retained throughout the evolution of both groups.

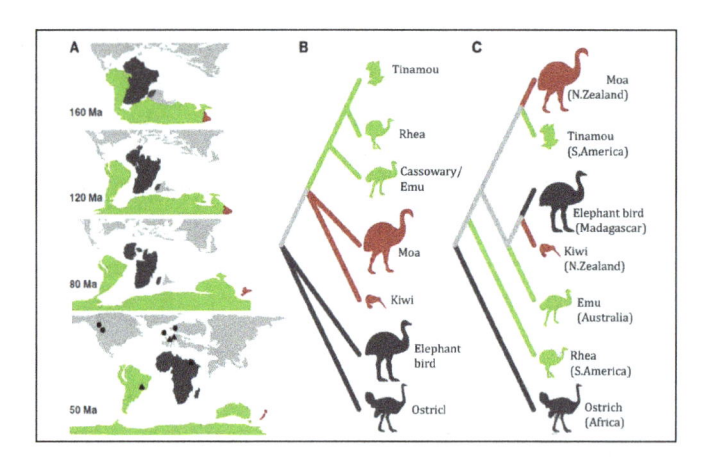

Flightless birds such as the emu, ostrich, and rhea fill very similar ecological niches on different continents. If ratites evolved from a Gondwanan common ancestor, they would not represent evolutionary convergence but instead would constitute an example of a shared (and conserved) ancestral flightless state. Now thought to be convergent, DNA evidence suggests that these "ratites" do not share a common ancestry but have evolved independently.

Live bearing, or viviparity, has evolved over 100 times among squamate reptiles (lizards and snakes), usually in response to cold climates. The probable mechanism behind the evolution of viviparity is that, by holding her eggs, a gravid female can both protect them from predators and, by basking, warm them, which would increase rate of development. Eventually, such selective forces favoring egg retention could lead to eggs hatching within a mother and live birth. This has happened even in geckos, all of which lay eggs except for one genus in New Caledonia and several related cold temperate New Zealand forms, which bear their young alive. In some skinks and xantusiid lizards, embryos attach to their mother's oviducts and grow, gaining nutrients during development via placental arrangements reminiscent of those in mammals.

Horned Lizard Phrynosoma cornutum   Thorny Devil Moloch horridus

Convergent evolutionary responses of lizards to arid environments are evident between continents. For example, Australian and North American deserts both support a cryptically-colored and thornily-armored ant specialized species: the thorny devil, Moloch horridus, an agamid, exploits this ecological role in Australia, while its counterpart the desert horned lizard (Phrynosoma), an iguanid, occupies it in North America. No Kalahari lizard has adopted such a life style. Interestingly, morphometric analysis demonstrates that the thorny devil and desert horned lizard are actually anatomically closer to one another than either species is to another member of its own lizard fauna, which are much more closely related.

Green tree python Chondropython viridis    Emerald Tree Boa Corallus caninus

Emerald Tree Boas from South American Amazonian rainforests are strikingly convergent with Green Tree Pythons found halfway around the world in similar rainforests in Australia, a spectacular example of ecological equivalents. Both of these snakes live high up in the canopy and eat birds. Adults are green, cryptically colored, matching the colors of leaves. Juveniles of both species are bright yellow or orange.

Juvenile Green Tree Python    A skink (Morethia butleri) from Australia

Juvenile Emerald Tree Boa    Blue-tailed anguid from Mexico (Celestes)

Colorful, blue, red and yellow tails have evolved repeatedly among distantly related lizards in many families (agamids, anguids, gymnopthlamids, lacerids, skinks, and teiids), presumably a ploy to attract a predator's attention away from the head to the tail, which can be broken off and regenerated should a predator attack it.

The New World iguanid Basiliscus, sometimes called 'Jesus lizards,' because they can run across the surface of water, have undergone convergent evolution with the Old World agamid Hydrosaurus. Both Basiliscus and Hydrosaurus have enlarged rectangular, plate-like, fringed scales on their toes, which allow these big lizards to run across water using surface tension for support.

Open sandy deserts pose severe problems for their inhabitants: (1) windblown sands are always loose and provide little traction; (2) surface temperatures at midday rise to lethal levels; and (3) open sandy areas offer little food or shade or shelter for evading predators. Even so, natural selection over eons of time has enabled lizards to cope fairly well with such sandy desert conditions. Subterranean lizards simply bypass most problems by staying underground, and actually benefit from the loose sand since underground locomotion is facilitated. Burrowing is also made easier by evolution of a pointed, shovel-shaped head and a countersunk lower jaw, as well as by small appendages and muscular bodies and short tails. Such a reduced-limb adaptive suite associated with fossorial habits has evolved repeatedly among squamate reptiles in both lizards and snakes.

During the hours shortly after sunrise, but before sand temperatures climb too high, diurnal lizards scurry about above ground in sandy desert habitats. Sand specialized lizards provide one of the most striking examples of convergent evolution and ecological equivalence. Representatives of many different families of lizards scattered throughout the world's deserts have found a similar solution for getting better traction on loose sand: enlarged scales on their toes, or lamellae, have evolved independently in six different families of lizards: skinks, lacertids, iguanids, agamids, gerrhosaurids, and geckos. A skink, Scincus, appropriately dubbed the 'sand fish,' literally swims through sandy seas in search of insect food in the Sahara and other eastern deserts. These sandy desert regions also support lacertid lizards (Acanthodactylus) with fringed toes and shovel noses. Far away in the southern hemisphere, on the windblown dunes of the Namib desert of southwestern Africa, an independent lineage of lacertids, Meroles (formerly Aporosaura) anchietae, has evolved a similar life form. In North America, this body form has been adopted by members of the iguanid genus Uma, which usually forage by waiting in the open and eat a fairly diverse diet of various insects, such as sand roaches, beetle larvae and other burrowing arthropods. They also listen intently for insects moving buried in the sand, and dig them up. Sometimes they dash, dig, and paw through a patch of sand and then watch the disturbed area for movements.

Meroles (Aporosaura) anchiete   Toes of Uma scoparia

All of these lizards have flattened, duckbill-like, shovel-nosed snouts, which enable them to make remarkable 'dives' into the sand even while running at full speed. The lizards then wriggle along under the surface, sometimes for over a meter. One must see such a sand diving act to appreciate fully its effectiveness as a disappearing act. Some Namib desert lizards have discovered another solution to gain traction on powdery sands: frog-like webbing between the toes as seen in the geckos Kaokogecko and Palmatogecko (now Pachydactylus).

Pachydactylus (formerlyPalmatogecko) rangeri

Other lizards that have undergone convergent evolution include rock mimics such as the North American horned lizard Phrynosoma modestum, an iguanid, and the Australian agamid Tympanocryptis cephalus. New World teiids (Tupinambis) have converged on Old World varanids (Varanus): members of both genera are large predatory lizards with forked tongues which they use as edge detectors to find scent trails and track down their prey (other vertebrates).

Rock Mimic Tympanocryptis cephalus.

Sometimes, roughly similar ecological systems support relatively few conspicuous ecological equivalents but instead are composed largely of distinctly different plant and animal types. For instance, although bird species diversities of temperate forests in eastern North America and eastern Australia are similar (Recher 1969), many avian niches appear to be fundamentally different on the two continents. Honeyeaters and parrots are conspicuous in Australia whereas hummingbirds and woodpeckers are entirely absent. Apparently different combinations of the various avian ecological activities are possible; thus, an Australian honeyeater might combine aspects of the food and place niches exploited in North America by both warblers and hummingbirds. An analogy can be made by comparing the 'total avian niche space' to a deck of cards. This niche space can be exploited in a limited number of ways, and each bird population or species has its own ways of doing things, or its own "hand of cards," determined in part by what other species in the community are doing.

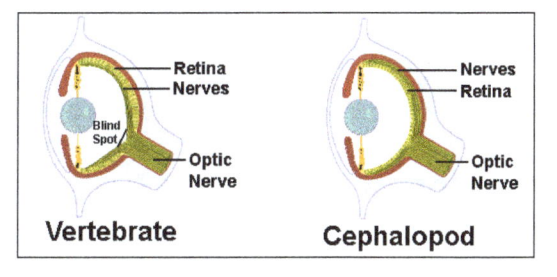

A very striking example of evolutionary convergence involves the eyes of vertebrates and cephalopod mollusks. Both have independently evolved complex camera-like eyes complete with an aperture, lens, and retina. Prominent anti-Darwinian Charles Hodges once suggested that the vertebrate eye was too complex to have evolved by natural selection and therefore must have been "designed." However, vertebrate eyes are poorly designed as compared to cephalopod eyes. In vertebrates, nerve fibers pass in front of the retina creating a blind spot, whereas nerves lie behind the retina in the superior cephalopod eye which does not have a blind spot.

Divergent Evolution

Divergent evolution is the process whereby groups from the same common ancestor evolve and accumulate differences, resulting in the formation of new species.

Divergent evolution may occur as a response to changes in abiotic factors, such as a change in environmental conditions, or when a new niche becomes available. Alternatively, divergent evolution may take place in response to changes in biotic factors, such as increased or decreased pressure from competition or predation.

As selective pressures are placed upon organisms, they must develop adaptive traits in order to survive and maintain their reproductive fitness. Differences may be minor, such as the change in shape, size or function of only one structure, or they may be more pronounced and numerous, resulting in a completely different body structure or phenotype.

Divergent evolution leads to speciation, and works on the basis that there is variation within the gene pool of a population. If a reproductive barrier separates two groups within a population, different genes controlling for various aspects of an organism's ability to survive and reproduce increase or decrease in frequency as gene flow is restricted. Allopatric speciation and peripatric speciation occur when the reproductive barrier is caused by a physical or geographical barrier, such as a river or mountain range. Alternatively, sympatric speciation and parapatric speciation take place within the same geographical area.

Through divergent evolution, organisms may develop homologous structures. These are anatomically similar structures, which are present in the common ancestor and persist within the diverged organisms, although have evolved dissimilar functions.

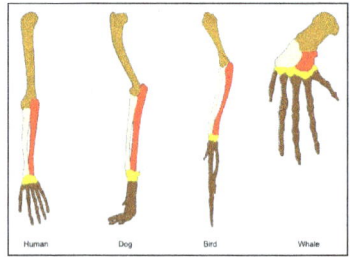

The image shows an example of the homologous bones found in the forelimb of four different types of mammal.

Examples of Divergent Evolution

Darwin's Finches

One of the most famous examples of divergent evolution was observed by Charles Darwin, and documented in his book On the Origin of Species.

Upon visiting the Galapagos Islands, Darwin noted that each of the islands had a resident population of finches belonging to the same taxonomic family. However, the bird populations on each island differed from those on nearby islands in the shape and size of their beaks.

Darwin suggested that each of the bird species had originally belonged to a single common ancestor species, which had undergone modifications of its features based on the type of food source available on each island. For example, the birds that fed on seeds and nuts evolved large crushing beaks, while cactus eaters developed longer beaks, and finer beaks evolved in birds that fed by picking insects out of trees.

When the ancestral form of finches initially colonized each island, each group contained individuals who were able to better adapt to the conditions and the available food source. These individuals survived and reproduced in their new habitat. In doing so, the genes that controlled for certain favorable aspects (e.g., longer beaks suitable for accessing nectar deep inside flowers) were spread throughout the gene pool, while the individuals without favored features died out. This is the process of natural selection.

The case of 'Darwin's Finches' (the birds actually belong to the tanager family and are not true finches) is an example of adaptive radiation, which is a form of divergent evolution.

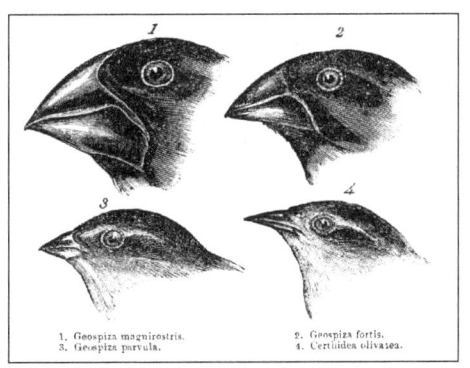

Darwin's finches

Adaptive radiation is a common feature in archipelagos such as the Galapagos Islands and Hawaii, as well as on metaphorical 'island habitats' such as mountain ranges. This is because gene flow between islands is limited when migration is not constant; however, the scale of the effect depends on the dispersal ability of the organism.

The Evolution of Primates

All of the primates on Earth evolved from a single common ancestor, most likely a primate-like, insectivore mammal, which lived around 65 million years ago in the Mesozoic Era. At that time, the world's continents were mostly connected. Fossil evidence suggests that these primitive animals lived an arboreal life, with good eyesight and hands and feet adapted to climbing through trees.

Around 55 million years ago, the first true primates evolved, diverging into the prosimians and simians.

Ancestral prosimians mostly resembled modern prosimians, which include the lemurs (endemic to Madagascar), lorises, tarsiers and bush babies. These are small-brained and relatively small-bodied, with a wet nose similar to that of a dog. They are often nocturnal, with body features that are considered 'primitive', compared to other primates.

The next big divergence occurred around 35 million years ago in the other phylogenetic branch of primates, the simians. This event resulted in the divergence of the common ancestor of all New World monkeys and Old World monkeys.

It is speculated that the two groups underwent divergent evolution as a consequence of allopatric speciation. As the continents of America and Eurasia had by this point separated, the split could have been caused by a chance migration across the Atlantic Ocean.

The New World monkeys or Platyrrhines, are native to Central and South America, as well as Mexico. They evolved flat noses and prehensile tails, which act as a fifth limb and have the ability to grasp on to trees and branches. These include familiar families such as capuchins and spider monkeys (family: Cebidae), marmosets (Callitrichidae), and howler monkeys (Atelidae).

The common ancestor of the Old World monkeys and apes split around 25 million years ago. Old World monkeys, or Catarrhini, are native to Africa and Asia, displaying a range of different adaptions to many types of habitat, from rainforests to savannah, mountains and shrubs. There are both terrestrial and arboreal Catarrhini, many of which are familiar, such as macaques genus: Macaca), baboons (Papio) and langurs (Semnopithecus).

The apes, or Hominoidea, further diverged into two groups: the lesser apes, such as gibbons (family: Hylobatidae), which are all native to Asia, and the great apes (Hominidae), which are native to Europe, Africa and Asia, and include orangutans (genus: Pongo), gorillas (Gorilla), chimpanzees (Pan) and humans (Homo).

It is important to remember that the modern primates we see today are not evolved from each other despite their similarities (for example, great apes are not evolved from lesser apes), but that they are descended from a single common ancestor that formed two different species through divergent evolution.

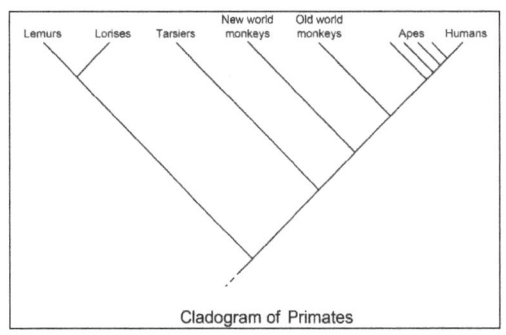

Primate cladogram

The Kit Fox and the Arctic Fox

Two species that are very closely related and have undergone divergent evolution are the kit fox (Vulpes macrotis) and the Arctic fox (Vulpes lagopus).

The kit fox is native to Western North America, and is adapted to desert environments; it has sandy coloration, and large ears, which help it to remove excess body heat.

The Arctic fox is native to Arctic regions and lives in the Arctic tundra biome of the Northern Hemisphere. Best adapted to cold climates, it has thick fur, which is white in the winter and brown in the summer, and a small, round body shape that minimizes heat loss.

Having diverged from a recent common ancestor, both these species have had to adapt to their extremely different habitats. They have evolved into two species that are clearly very distinct in terms of their ears and coats, although they still retain the majority of their ancestral features.

Parallel Evolution

Parallel evolution refers to the evolutionary process wherein two or more species in the same environment develop similar adaptation or characteristics. Example of parallel evolution: North American cactus and the African euphorbia that developed similar adaptation, which is their thick stems and sharp quills to survive the hot, arid climates. These two plant species are of different plant families but live in the same type of environment. Another example is the evolution of adaptive features between two groups of organisms living in similar habitats such as marsupial mammals in Australia and placental mammals on another continent.

Parallel evolution is similar to convergent evolution in a way that two unrelated species evolved similar traits. However, in parallel evolution, the two species evolved same traits while living in the same type of environment whereas in convergent evolution the two species evolved same traits in different types of environment

Theories of Evolution

The main theories of evolution are:
- Lamarckism or Theory of Inheritance of Acquired characters.
- Darwinism or Theory of Natural Selection.
- Mutation theory of De Vries.
- Neo-Darwinism or Modern concept or Synthetic theory of evolution.

Lamarckism

Figure: Jean Baptiste de Lamarck.

It is also called "Theory of inheritance of acquired characters" and was proposed by a great French naturalist, Jean Baptiste de Lamarck in 1809 A.D. in his famous book "Philosphic Zoologique". This theory is based on the comparison between the contemporary species of his time to fossil records.

His theory is based on the inheritance of acquired characters which are defined as the changes (variations) developed in the body of an organism from normal characters, in response to the changes in environment, or in the functioning (use and disuse) of organs, in their own life time, to fulfill their new needs. Thus Lamarck stressed on adaptation as means of evolutionary modification.

Postulates of Lamarckism

Lamarckism is based on following four postulates:

1. New needs:

Every living organism is found in some kind of environment. The changes in the environmental factors like light, temperature, medium, food, air etc. or migration of animal lead to the origin of new needs in the living organisms, especially animals. To fulfill these new needs, the living organisms have to exert special efforts like the changes in habits or behaviour.

2. Use and disuse of organs:

The new habits involve the greater use of certain organs to meet new needs, and the disuse or lesser use of certain other organs which are of no use in new conditions. This use and disuse of organs greatly affect the form, structure and functioning of the organs.

Continuous and extra use of organs make them more efficient while the continued disuse of some other organs lead to their degeneration and ultimate disappearance. So, Lamarckism is also called "Theory of use and disuse of organs."

So the organism acquires certain new characters due to direct or indirect environmental effects during its own life span and are called Acquired or adaptive characters.

3. Inheritance of acquired characters:

Lamarck believed that acquired characters are inheritable and are transmitted to the offsprings so that these are born fit to face the changed environmental conditions and the chances of their survival are increased.

4. Speciation:

Lamarck believed that in every generation, new characters are acquired and transmitted to next generation, so that new characters accumulate generation after generation. After a number of generations, a new species is formed.

So according to Lamarck, an existing individual is the sum total of the characters acquired by a number of previous generations and the speciation is a gradual process.

Four Postulates of Lamarckism

1. Living organisms or their component parts tend to increase in size.

2. Production of new organ is resulted from a new need.

3. Continued use of an organ makes it more developed, while disuse of an organ results in degeneration.

4. Acquired characters (or modifications) developed by individuals during their own lifetime are inheritable and accumulate over a period of time resulting a new species.

Evidences in Favour of Lamarckism

1. Phylogenetic studies of horse, elephant and other animals show that all these increase in their evolution from simple to complex forms.

2. Giraffe:

Development of present day long-necked and long fore-necked giraffe from deer-like ancestor by the gradual elongation of neck and forelimbs in response to deficiency of food on the barren ground in dry deserts of Africa. These body parts were elongated so as to eat the leaves on the tree branches. This is an example of effect of extra use and elongation of certain organs.

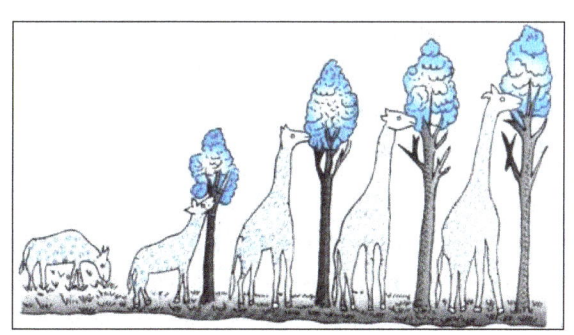

Figure: Stages in the evolution of present-day giraffe.

3. Snakes:

Development of present day limbless snakes with long slender body from the limbed ancestors due to continued disuse of limbs and stretching of their body to suit their creeping mode of locomotion and fossorial mode of living out of fear of larger and more powerful mammals. It is an example of disuse and degeneration of certain organs.

4. Aquatic birds:

Development of aquatic birds like ducks, geese etc. from their terrestrial ancestors by the acquired characters like reduction of wings due to their continued disuse, development of webs between their toes for wading purposes.

These changes were induced due to deficiency of food on land and severe competition. It is an example of both extra use (skin between the toes) and disuse (wings) of organs.

5. Flightless birds:

Development of flightless birds like ostrich from flying ancestors due to continued disuse of wings as these were found in well protected areas with plenty of food.

6. Horse:

The ancestors of modem horse (Equus caballus) used to live in the areas with soft ground and were short legged with more number of functional digits (e.g. 4 functional fingers and 3 functional toes in Dawn horse-Eohippus). These gradually took to live in areas with dry ground. This change in habit was accompanied by increase in length of legs and decrease in functional digits for fast running over hard ground.

Criticism of Lamarckism

A hard blow to Lamarckism came from a German biologist, August Weismann who proposed the "Theory of continuity of germplasm" in 1892 A.D. This theory states that environmental factors do affect only somatic cells and not the germ cells.

As the link between the generations is only through the germ cells and the somatic cells are not transmitted to the next generation so the acquired characters must be lost with the death of an organism so these should have no role in evolution. He suggested that germplasm is with special particles called "ids" which control the development of parental characters in offsprings.

Weismann mutilated the tails of mice for about 22 generations and allowed them to breed, but tailless mice were never born. Pavlov, a Russian physiologist, trained mice to come for food on hearing a bell. He reported that this training is not inherited and was necessary in every generation. Mendel's laws of inheritance also object the postulate of inheritance of acquired characters of Lamarckism.

Similarly, boring of pinna of external ear and nose in Indian women; tight waist, of European ladies circumcising (removal of prepuce) in certain people; small sized feet of Chinese women etc are not transmitted from one generation to another generator.

Eyes which are being used continuously and constantly develop defects instead of being improved. Similarly, heart size does not increase generation after generation though it is used continuously.

Presence of weak muscles in the son of a wrestler was also not explained by Lamarck. Finally, there are a number of examples in which there is reduction in the size of organs e.g. among Angiosperms, shrubs and herbs have evolved from the trees.

So, Lamarckism was rejected.

Significance

1. It was first comprehensive theory of biological evolution.
2. It stressed on adaptation to the environment as a primary product of evolution.

Neo-Lamarckism

Long forgotten Lamarckism has been revived as Neo-Lamarckism, in the light of recent findings in the field of genetics which confirm that environment does affect the form, structure; colour, size etc. and these characters are inheritable.

Main scientists who contributed in the evolution of Neo-Lamarckism are: French Giard, American Cope, T.H. Morgan, Spencer, Packard, Bonner, Tower, Naegali, Mc Dougal, etc. Term neo-Lamarckism was coined by Alphaeus S. Packard.

Neo-Lamarckism States

1. Germ cells may be formed from the somatic cells indicating similar nature of chromosomes and gene make up in two cell lines e.g.

(a) Regeneration in earthworms.

(b) Vegetative propagation in plants like Bryophyllum (with foliar buds).

(c) A part of zygote (equipotential egg) of human female can develop into a complete baby (Driesch).

2. Effect of environment on germ cells through the somatic cells e.g. Heslop Harrison found that a pale variety of moth, Selenia bilunaria, when fed on manganese coated food, a true breeding melanic variety of moth is produced.

3. Effect of environment directly on germ cells. Tower exposed the young ones of some potato beetles to temperature fluctuation and found that though beetles remained unaffected with no somatic change but next generation had marked changes in body colouration.

Muller confirmed the mutagenic role of X-rays on Drosophila while C. Auerbach et., al. confirmed the chemical mutagens (mustard gas vapours) causing mutation in Drosophila melanogaster, so neo-Lamarckism proved:

(a) Germ cells are not immune from the effect of environment.

(b) Germ cells can carry somatic changes to next progeny (Harrison's experiment).

(c) Germ cells may be directly affected by the environmental factors (Tower's experiment).

Differences between Lamarckism and Neo-Lamarckism.

Character	Lamarckism	Neo-Lamarckism
1. Nature of theory	Original as proposed by Lamarck.	Modified Lamarckism in the light of modern knowledge.
2. Factors inducing variations	Certain internal forces, changes environmental factors and use and disuse of organs.	Changes in environmental factors but not due to use and disuse of organs and internal forces.
3. Cells involved	Only somatic cells are affected so acquired characters are developed during individuals own life span.	Somatic cells or Germ cells or both are affected.
4. Nature of inherited characters	Acquired characters are inheritable.	Only germinal variations are inheritable or where germ cells are formed from somatic cells.

Darwinism (Theory of Natural Selection)

Charles Darwin an English naturalist, was the most dominant figure among the biologists of the 19th century. He made an extensive study of nature for over 20 years, especially in 1831-1836 when he went on a voyage on the famous ship "H.M.S. Beagle" and explored South America, the Galapagos Islands and other islands.

Figure: Charles Robert Darwin.

He collected the observations on animal distribution and the relationship between living and extinct animals. He found that existing living forms share similarities to varying degrees not only among themselves but also with the life forms that existed millions of years ago, some of which have become extinct.

Figure: H.M.S. Beagle ship.

He stated that every population has built in variations in their characters. From the analysis of his data of collection and from Malthus's Essay on Population, he got the idea of struggle for existence within all the populations due to continued reproductive pressure and limited resources and that all organisms, including humans, are modified descendents of previously existing forms of life.

In 1858 A.D., Darwin was highly influenced by a short essay entitled "On the Tendency of Varieties to Depart Indefinitely from the Original Type" written by another naturalist, Alfred Russel Wallace (1812-1913) who studied biodiversity on Malayan archipelago and came to similar conclusions.

Darwin and Wallace's views about evolution were presented in the meeting of Linnean Society of London by Lyell and Hooker on July 1, 1858. Darwin's and Wallace's work was jointly published in "Proceedings of Linnean Society of London" in 1859. So it is also called Darwin-Wallace theory.

Darwin explained his theory of evolution in a book entitled "On the Origin of Species by means of Natural Selection". It was published on 24th Nov., 1859. In this theory, Charles Darwin proposed the concept of natural selection as the mechanism of evolution.

Postulates of Darwinism

Main postulates of Darwinism are:

1. Geometric increase.
2. Limited food and space.
3. Struggle for existence.
4. Variations.
5. Natural selection or Survival of the fittest.
6. Inheritance of useful variations.
7. Speciation.

Geometric Increase

According to Darwinism, the populations tend to multiply geometrically and the reproductive powers of living organisms (biotic potential) are much more than required to maintain their number e.g.,

Paramecium divides three times by binary fission in 24 hours during favourable conditions. At this rate, a Paramecium can produce a clone of about 280 million Paramecia in just one month and in five years, can produce Paramecia having mass equal to 10,000 times than the size of the earth.

Other rapidly multiplying organisms are: Cod (one million eggs per year); Oyster (114 million eggs in one spawning); Ascaris (70, 00,000 eggs in 24 hours); housefly (120 eggs in one laying and laying eggs six times in a summer season); a rabbit (produces 6 young ones in a litter and four litters in a year and young ones start breeding at the age of six months).

Similarly, the plants also reproduce very rapidly e.g., a single evening primrose plant produces about 1, 18,000 Seeds and single fern plant produces a few million spores.

Even slow breeding organisms reproduce at a rate which is much higher than required e.g., an elephant becomes sexually mature at 30 years of age and during its life span of 90 years, produces only six offsprings. At this rate, if all elephants survive then a single pair of elephants can produce about 19 million elephants in 750 years.

These examples confirm that every species can increase manifold within a few generations and occupy all the available space on the earth, provided all survive and repeat the process. So the number of a species will be much more than can be supported on the earth.

Limited Food and Space

Darwinism states that though a population tends to increase geometrically, the food increases only arithmetically. So two main limiting factors on the tremendous increase of a population are: limited food and space which together form the major part of carrying capacity of environment. These do not allow a population to grow indefinitely which are nearly stable in size except for seasonal fluctuation.

Struggle for Existence

Due to rapid multiplication of populations but limited food and space, there starts an everlasting competition between individuals having similar requirements. In this competition, every living organism desires to have an upper hand over others.

This competition between living organisms for the basic needs of life like food, space, mate etc., is called struggle for existence which is of three types:

(a) Intraspecific:

Between the members of same species e.g. two dogs struggling for a piece of meat.

(b) Interspecific:

Between the members of different species e.g. between predator and prey.

(c) Environmental or Extra specific:

Between living organisms and adverse environmental factors like heat, cold, drought, flood, earthquakes, light etc.

Out of these three forms of struggle, the intraspecific struggle is the strongest type of struggle as the needs of the individuals of same species are most similar e.g., sexual selection in which a cock with a more beautiful comb and plumage has better chances to win a hen than a cock with less developed comb.

Similarly, cannabilism is another example of intraspecific competition as in this; individuals eat upon the members of same species.

In this death and life struggle, the majority of individuals die before reaching the sexual maturity and only a few individuals survive and reach the reproductive stage. So struggle for existence acts as an effective check on an ever-increasing population of each species.

The nature appears saying, "They are weighed in the balance and are found wanting." So the number of offsprings of each species remains nearly constant over long period of time.

Variations

Variation is the law of nature. According to this law of nature, no two individuals except identical (monozygotic) twins are identical. This everlasting competition among the organisms has compelled them to change according to the conditions to utilize the natural resources and can survive successfully.

Darwin stated that the variations are generally of two types—continuous variations or fluctuations and discontinuous variations. On the basis of their effect on the survival chances of living organisms, the variations may be neutral, harmful and useful.

Darwin proposed that living organisms tend to adapt to changing environment due to useful continuous variations {e.g., increased speed in the prey; increased water conservation in plants; etc.), as these will have a competitive advantage.

Natural Selection or Survival of the Fittest

Darwin stated that as many selects the individuals with desired characters in artificial selection; nature selects only those individuals out of the population which are with useful continuous variations and are best adapted to the environment while the less fit or unfit individuals are rejected by it.

Darwin stated that if the man can produce such a large number of new species/varieties with limited resources and in short period of time by artificial selection, then natural selection could account for this large biodiversity by considerable modifications of species with the help of unlimited resources available over long span of time.

Darwin stated that discontinuous variations appear suddenly and will mostly be harmful, so are not selected by nature. He called them "sports". So the natural selection is an automatic and self going process and keeps a check on the animal population.

This sorting out of the individuals with useful variations from a heterogeneous population by the nature was called Natural selection by Darwin and Survival of the fittest by Wallace. So natural selection acts as a restrictive force and not a creative force.

Inheritance of useful Variations

Darwin believed that the selected individuals pass their useful continuous variations to their offsprings so that they are born fit to the changed environment.

Speciation

According to Darwinism, useful variations appear in every generation and are inherited from one generation to another. So the useful variations go on accumulating and after a number of generations, the variations become so prominent that the individual turns into a new species. So according to Darwinism, evolution is a gradual process and speciation occurs by gradual changes in the existing species.

Thus the two key concepts of Darwinian Theory of Evolution are:

1. Branching Descent, and 2. Natural Selection.

Evidences in Favour of Darwinism

1. There is a close parallelism between natural selection and artificial selection.

2. The remarkable cases of resemblance e.g. mimicry and protective colouration can be achieved only by gradual changes occurring simultaneously both in the model and the mimic.

3. Correlation between position of nectaries in the flowers and length of the proboscis of the pollinating insect.

Evidences against Darwinism

Darwinism is not Able to explain:

1. The inheritance of small variations in those organs which can be of use only when fully formed e.g. wing of a bird. Such organs will be of no use in incipient or underdeveloped stage.

2. Inheritance of vestigial organs.

3. Inheritance of over-specialised organs e.g. antlers in deer and tusks in elephants.

4. Presence of neuter flowers and sterility of hybrids.

5. Did not differentiate between somatic and germinal variations.

6. He did not explain the causes of the variations and the mode of transmission of variations.

7. It was also refuted by Mendel's laws of inheritance which state that inheritance is particulate.

So this theory explains only the survival of the fittest but does not explain the arrival of the fittest so Darwin himself confessed, "natural selection has been main but not the exclusive means of modification."

Principle of Natural Selection

It was proposed by Ernst Mayer in 1982. It stems from five important observations and three inferences as shown in table below. This principle demonstrates that natural selection is the differential success in reproduction and enables the organisms to adapt them to their environment by development of small and useful variations.

Observations	Inferences
1. All species have such great potential of fertility that their population size would increase exponentially if all individuals that were bron reproduced successfully. 2. Most populations are normally stable in size, except for seasonal fluctuations. 3. Natural resources are limited. 4. Individuals of a population vary extensively in their characteristics; no two individuals are exactly alike. 5. Much of this variation is heritable.	(a) Production of more individuals than the environment can support leads to a struggle for existence among individuals of a population, with only a fraction of offspring surviving each generation. (b) Survival in the struggle for existence is not random, but depends in part on the hereditary constitution of the surviving individuals. Those individuals whose inherited characteristics fit them best in their environment are likely to leave more offsprings than less fit individuals. (c) The unequal ability of individuals to survive and reproduce will lead to a gradual change in a population with favourable characteristics accumulating over the generations.

These favourable Variations accumulate over generation after generation and lead to speciation. So natural selection operates through interactions between the environment and inherent variability in the population.

Mutation Theory of Evolution

The mutation theory of evolution was proposed by a Dutch botanist, Hugo de Vries. He worked on evening primrose (Oenothera lamarckiana).

Figure: Hugo de Vries.

Experiment

Hugo de Vries cultured O. lamarckiana in botanical gardens at Amsterdam. The plants were, allowed to self pollinate and next generation was obtained. The plants of next generation were again subjected to self pollination to obtain second generation. Process was repeated for a number of generations.

Observations

Majority of plants of first generation were found to be like the parental type and showed only minor variations but 837 out of 54,343 members were found to be very different in characters like flower size, shape and arrangement of buds, size of seeds etc. These markedly different plants were called primary or elementary species.

A few plants of second generation were found to be still more different. Finally, a new type, much longer than the original type, called O. gigas, was produced. He also found the numerical chromosomal changes in the variants (e.g. with chromosome numbers 16, 20, 22, 24, 28 and 30) upto 30 (Normal diploid number is 14).

1. The evolution is a discontinuous process and occurs by mutations (L. mutate = to change; sudden and inheritable large differences from the normal and are not connected to normal by intermediate forms). Individuals with mutations are called mutants.

2. Elementary species are produced in large number to increase chances of selection by nature.

3. Mutations are recurring so that the same mutants appear again and again. This increases the chances of their selection by nature.

4. Mutations occur in all directions so may cause gain or loss of any character.

5. Mutability is fundamentally different from fluctuations (small and directional changes).

So according to mutation theory, evolution is a discontinuous and jerky process in which there is a jump from one species to another so that new species arises from pre-existing species in a single generation (macrogenesis or saltation) and not a gradual process as proposed by Lamarck and Darwin.

Evidences in Favour of Mutation Theory

1. Appearance of a short-legged sheep variety, Ancon sheep, from long-legged parents in a single generation in 1791 A.D. It was first noticed in a ram (male sheep) by an American farmer, Seth Wright.

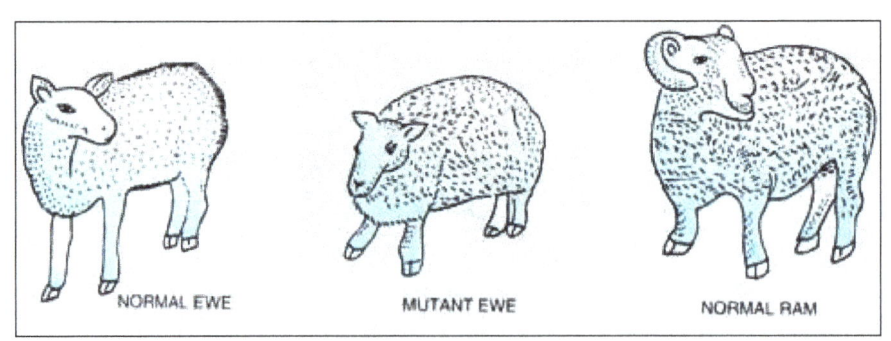

Figure: Appearance of short-legged Ancon sheep mutant.

2. Appearance of polled Hereford cattle from horned parents in a single generation in 1889.

3. De Vries observations have been experimentally confirmed by McDougal and Shull in America and Gates in England.

4. Mutation theory can explain the origin of new varieties or species by a single gene mutation e.g. Cicer gigas, Nuval orange. Red sunflower, hairless cats, double- toed cats, etc.

5. It can explain the inheritance of vestigial and over-specialized organs.

6. It can explain progressive as well as retrogressive evolution.

Evidences against Mutation Theory

1. It is not able to explain the phenomena of mimicry and protective colouration.

2. Rate of mutation is very low, i.e. one per million or one per several million genes.

3. Oenothera lamarckiana is a hybrid plant and contains anamolous type of chromosome behaviour.

4. Chromosomal numerical changes as reported by de Vries are unstable.

5. Mutations are incapable of introducing new genes and alleles into a gene pool.

Neo-Darwinism or Modern Concept or Synthetic Theory of Evolution

The detailed studies of Lamarckism, Darwinism and Mutation theory of evolution showed that no single theory is fully satisfactory. Neo-Darwinism is a modified version of theory of Natural Selection and is a sort of reconciliation between Darwin's and de Vries theories.

Modern or synthetic theory of evolution was designated by Huxley (1942). It emphasises the importance of populations as the units of evolution and the central role of natural selection as the most important mechanism of evolution.

The scientists who contributed to the outcome of Neo-Darwinism were: J.S. Huxley, R.A. Fischer and J.B.S. Haldane of England; and S. Wright, Ford, H.J. Muller and T. Dobzhansky of America.

Postulates of Neo-Darwinism

Genetic Variability

Variability is an opposing force to heredity and is essential for evolution as the variations form the raw material for evolution. The studies showed that the units of both heredity and mutations are genes which are located in a linear manner on the chromosomes.

Various sources of genetic variability in a gene pool are:

Mutations

These are sudden, large and inheritable changes in the genetic material. On the basis of amount of genetic material involved, mutations are of three types:

Chromosomal aberrations

These include the morphological changes in the chromosomes without affecting the number of chromosomes. These result changes either in the number of genes (deletion and duplication) or in the position of genes (inversion).

These are of four Types:

1. Deletion (Deficiency) involves the loss of a gene block from the chromosome and may be terminal or intercalary.

2. Duplication involves the presence of some genes more than once, called the repeat. It may be tandem or reverse duplication.

3. Translocation involves transfer of a gene block from one chromosome to a non-homologous chromosome and may be simple or reciprocal type.

4. Inversion involves the rotation of an intercalary gene block through 180° and may be paracentric or pericentric.

Numerical Chromosomal Mutations

These include changes in the number of chromosomes. These may be euploidy (gain or loss of one or more genomes) or aneuploidy (gain or loss of one or two chromosomes). Euploidy may be haploidy or polyploidy.

Among polyploidy, tetraploidy is most common. Polyploidy provides greater genetic material for mutations and variability. In haploids, recessive genes express in the same generation.

Aneuploidy may be hypoploidy or hyperploidyl Hypoploidy may be monosomy (loss of one chromosome) or nullisomy (loss of two chromosomes). Hyperploidy may be trisomy (gain of one chromosome) or tetrasomy (gain of two chromosomes).

Gene Mutations (Point Mutations)

These are invisible changes in chemical nature (DNA) of a gene and are of three types:

- Deletion involves loss of one or more nucleotide pairs.
- Addition involves gain of one or more nucleotide pairs.
- Substitution involves replacement of one or more nucleotide pairs by other base pairs. These may be transition or transversion type.

These changes in DNA cause the changes in the sequence of amino acids so changing the nature of proteins and the phenotype.

- Recombination of genes: Thousands of new combinations of genes are produced due to crossing over, chance arrangement of bivalents at the equator during metaphase – I and chance fusion of gametes during fertilization.
- Hybridization: It involves the interbreeding of two genetically different individuals to produce 'hybrids'.
- Physical mutagens (e.g. radiations, temperature etc.) and chemical mutagens (e.g. nitrous acid, colchicine, nitrogen mustard etc.).
- Genetic drift: It is the elimination of the genes of some original characteristics of a species by extreme reduction in a population due to epidemics or migration or Sewell Wright effect.

The chances of variations are also increased by non-random mating.

Natural Selection

Natural selection of Neo-Darwinism differs from that of Darwinism that it does not operate through "survival of the fittest" but operates through differential reproduction and comparative reproductive success.

Differential reproduction states that those members, which are best adapted to the environment, reproduce at a higher rate and produce more offsprings than those which are less adapted. So these contribute proportionately greater percentage of genes to the gene pool of next generation while less adapted individuals produce fewer offsprings.

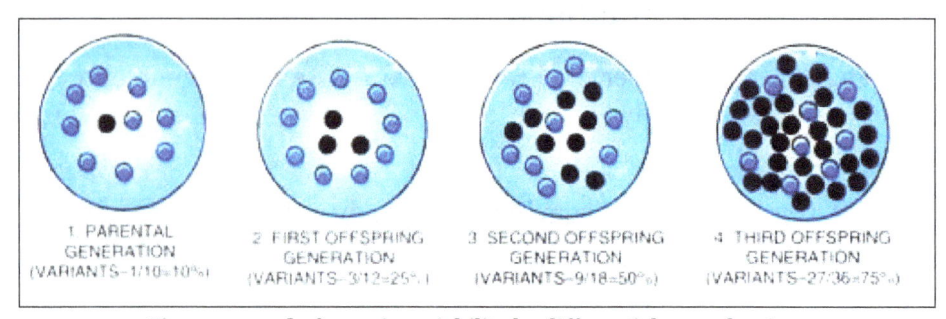

Figure: spread of genetic variability by differential reproduction.

If the differential reproduction continues for a number of generations, then the genes of those individuals which produce more offsprings will become predominant in the gene pool of the population.

Due to sexual communication, there is free flow of genes so that the genetic variability which appears in certain individuals, gradually spreads from one deme to another deme, from deme to population and then on neighbouring sister populations and finally on most of the members of a species. So natural selection causes progressive changes in gene frequencies, 'i.e. the frequency of some genes increases while the frequency of some other genes decreases.

Offsprings Produced by Individuals

(i) Mostly those individuals which are best adapted to the environment.

(ii) Whose sum of the positive selection pressure due to useful genetic variability is more than the sum of negative selection pressure due to harmful genetic variability?

(iii) Which have better chances of sexual selection due to development of some bright coloured spots on their body e.g. in many male birds and fish.

(iv) Those who are able to overcome the physical and biological environmental factors to successfully reach the sexual maturity.

So natural selection of Neo-Darwinism acts as a creative force and operates through comparative reproductive success. Accumulation of a number of such variations leads to the origin of a new species.

Reproductive Isolation

Any factor which reduces the chances of interbreeding between the related groups of living organisms is called an isolating mechanism. Reproductive isolation is must so as to allow the accumulation of variations leading to speciation by preventing hybridization.

In the absence of reproductive isolation, these variants freely interbreed which lead to intermixing of their genotypes, dilution of their peculiarities and disappearance of differences between them. So, reproductive isolation helps in evolutionary divergence.

Common Descent

Common descent is a term within evolutionary biology which refers to the common ancestry of a particular group of organisms. The process of common decent involves the formation of new species from an ancestral population. When a recent common ancestor is shared between two organisms, they are said to be closely related. In contrast, common descent can also be traced back to a universal common ancestor of all living organisms using molecular genetic methods. Such evolution from a universal common ancestor is thought to have involved several speciation events as a result of natural selection and other processes, such as geographical separation.

Theory of Common Descent

The theory of common descent states that all living organisms are descendants of a single ancestor. Thus, the Theory of Common Descent helps to explain why species living in different geographical regions exhibit different traits, some traits are highly conserved among broad animal classifications (e.g., vertebrates or tetrapods), seemly different species (e.g., birds and reptiles) share inherited physical and genetic traits, and successfully adapted organisms typically produce more offspring. While the Theory of Common Descent is primarily derived from the physical observation of various phenotypes (e.g., size, colour, beak shape, embryological development, etc.), modern advances in genetics and associated molecular techniques have been able to demonstrate that the process by which DNA is eventually translated into proteins is maintained among all lifeforms. Small changes in DNA between organisms have revealed a shared ancestry as well as insight into important changes that resulted in various speciation events. Phylogenetic trees and cladograms are often used to hypothesize the evolution of various organisms and shared common descent.

Examples of Common Descent

Human Chromosome 2

Compelling evidence of the shared common ancestry of humans with the great apes is the fusion event which occurred when two chromosomes common in apes fused to form chromosome 2 in humans (as illustrated below). This resulted in humans having only 23 pairs of chromosomes, while all other hominidae have 24 pairs. The great apes (e.g., chimpanzees, gorillas, and orangutans) have two chromosomes with almost identical DNA sequences as that found in chromosome 2 of humans. Further evidence of such a fusion event is the residual presence of telomeres and a centromere, which indicate that the genetic information was historically found on two separate chromosomes.

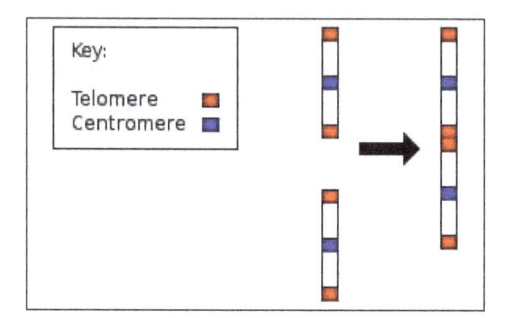

Endogenous Retroviruses

Endogenous retroviruses are residual DNA sequences found in the genomes of virtually all living organisms as a result of ancient viral infections. Since the retroviral sequences are incorporated into the DNA of the host organism, such sequences are inherited in the offspring. Since such infections are random events, as is the location in which the viral genome is inserted within the genome, the identification of the same retroviral sequences in multiple species is indicative of a shared ancestry. Such analyses of endogenous retroviruses often reveal speciation events (e.g., feline endogenous retroviruses reveal the separation between large and small cat species) and how closely related two species may be, as observed in the shared endogenous retroviruses between humans and other primate species.

The Presence of Atavisms

Atavisms are the appearance of a lost trait observed in an ancestral species that is not observed in more recent ancestors. Atavisms are an example of common descent as they provide evidence of the phenotypical or vestigial features that are often retained throughout evolution. Some examples, include the appearance of hind limbs in whales as evidence of a terrestrial ancestor, teeth exhibited by chickens, additional toes observed in modern horse species, and the back flippers of bottlenose dolphins. Atavisms tend to arise because the ancestral genes are not deleted from the genome, but rather silenced and then reactivated in later offspring.

Vestigial Structures

Similar to atavisms, vestigial structures are structures that are homologous to those found in ancestral species, but have become underdeveloped, non-functional, or degenerated in more recently evolved organisms. Such structures provide evidence of adaptations to a new environment, in which the ancestral organ or limb is no longer required, or has been modified to better suit a new purpose. There are an abundance of examples of vestigial structures observed in nature. Some examples include, the hind limbs and pelvic girdle observed in whales (as shown below) and snakes, non-functional wings exhibited by some insect species, non-functional wings of flightless bird species (e.g., ostriches), abdominal segments of barnacles, and the embryonic limb buds exhibited by several species lacking hind limbs (e.g., dolphins).

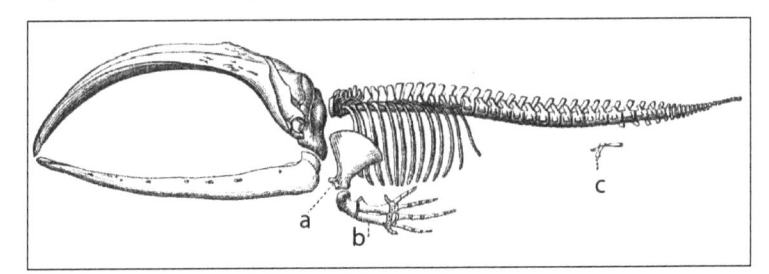

Pentadactyl Limbs

The presence of pentadactyl limbs is an example of a homologous trait exhibited by all tetrapods and is highly conserved throughout evolution, despite some modifications. Such limbs are first observed in the evolution from fishes to amphibians and consists of a single proximal, two distal, carpal, five metacarpal, and phalange bones. Although the overarching structure of the pentadactyl limbs is similar, various modifications have been made throughout evolution as adaptations to specific environments or lifestyles. Some examples include the modified pentadactyl wings of bats, the elongated forearms of primate species, the flippers of dolphin and whale species, and the modified digits of horses to form a hoof.

Fossil Evidence

The fossilized remains of various organisms combined with modern dating methods provides some of the most convincing evidence of common descent and evolutionary history. Fossilization occurs when the bones of a decaying animal become porous and the mineral salts in the surrounding earth infiltrate the bones, converting them to stone. Other methods include preservation in ice, imprinted remains (e.g., plants or footprints), tree resin, and peat. Since fossils are found in sedimentary rock, which is formed by layers of silt and mud, each layer corresponds to a specific geological period which can be dated. Thus, the extinction, evolution, and emergence of various species can be observed throughout history using the fossil record. Many extinct species are also observed in the fossil record, such as dinosaurs.

Biogeography

Biogeography presents compelling evidence of common descent by showcasing speciation and novel traits through adaptations to environmental pressures. One of the most famous examples is that of island biogeography and Charles Darwin's observations of the beaks of finches residing on the Galapagos Islands. In these finches, the beaks had been adapted for the specific vegetation found on the island, resulting in deviation from the ancestral finches found on the mainland. Long term effects of geographical separation are also observed with the evolution of new species that are not found elsewhere in the world. An example of this is the presence of marsupial species on the continent of Australia, and the emergence of polar bears as a result of geographical isolation in the arctic.

Speciation

New species arise through a process called speciation. In speciation, an ancestral species splits into two or more descendant species that are genetically different from one another and can no longer interbreed.

Darwin envisioned speciation as a branching event. In fact, he considered it so important that he depicted it in the only illustration of his famous book, On the Origin of Species,. A modern representation of Darwin's idea is shown in the evolutionary tree of elephants and their relatives, below right, which reconstructs speciation events during the evolution of this group.

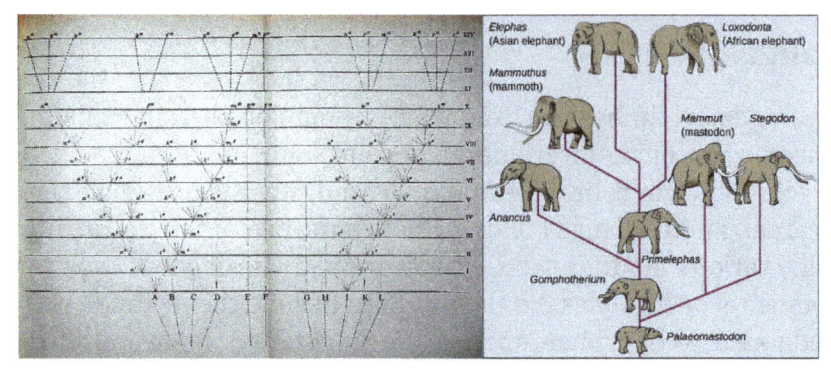

Formation of new species.

For speciation to occur, two new populations must be formed from one original population, and they must evolve in such a way that it becomes impossible for individuals from the two new populations to interbreed. Biologists often divide the ways that speciation can occur into two broad categories:

- Allopatric speciation: Allo meaning other and patric meaning homeland—involves geographic separation of populations from a parent species and subsequent evolution.

- Sympatric speciation: Sym meaning same and patric meaning homeland—involves speciation occurring within a parent species remaining in one location.

Allopatric Speciation

In allopatric speciation, organisms of an ancestral species evolve into two or more descendant species after a period of physical separation caused by a geographic barrier, such as a mountain range, rockslide, or river.

Sometimes barriers, such as a lava flow, split populations by changing the landscape. Other times, populations become separated after some members cross a pre-existing barrier. For example, members of a mainland population may become isolated on an island if they float over on a piece of debris.

Once the groups are reproductively isolated, they may undergo genetic divergence. That is, they may gradually become more and more different in their genetic makeup and heritable features over many generations. Genetic divergence happens because of natural selection, which may favor different traits in each environment, and other evolutionary forces like genetic drift.

As they diverge, the groups may evolve traits that act as prezygotic and/or postzygotic barriers to reproduction. For instance, if one group evolves large body size and the other evolves small body size, the organisms may not be physically able to mate—a prezygotic barrier—if the populations are reunited.

If the reproductive barriers that have arisen are strong—effectively preventing gene flow—the groups will keep evolving along separate paths. That is, they won't exchange genes with one another even if the geographical barrier is removed. At this point, the groups can be considered separate species.

Sympatric Speciation

In sympatric speciation, organisms from the same ancestral species become reproductively isolated and diverge without any physical separation.

At first, this idea may seem kind of weird, especially after thinking about allopatric speciation. Why would groups of organisms in a population stop interbreeding when they still live in the same place?

There are several ways that sympatric speciation can happen. However, one mechanism that's quite common—in plants, that is!—involves chromosome separation errors during cell division. Let's take a closer look at this process.

Polyploidy

Polyploidy is the condition of having more than two full sets of chromosomes. Unlike humans and other animals, plants are often tolerant of changes in their number of chromosome sets, and an increase in chromosome sets, a.k.a. ploidy, can be an instant recipe for plant sympatric speciation.

How could polyploidy lead to speciation? As an example, let's consider the case where a tetraploid plant—4n, having four chromosome sets—suddenly pops up in a diploid population—2n, having two chromosome sets.

Such a tetraploid plant might arise if chromosome separation errors in meiosis produced a diploid egg and a diploid sperm that then met up to make a tetraploid zygote. This process is shown in a general schematic below.

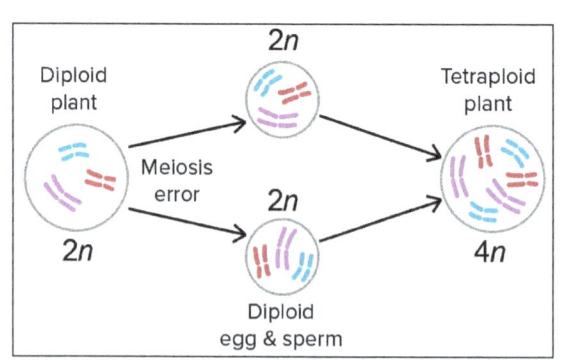

When the tetraploid plant matures, it will make diploid, 2n, eggs and sperm. These eggs and sperm can readily combine with other diploid eggs and sperm via self-fertilization, which is common in plants, to make more tetraploids.

On the other hand, the diploid eggs and sperm may or may not combine effectively with the haploid, 1n, eggs and sperm from the parental species. Even if the diploid and haploid gametes do get together to produce a triploid plant with three chromosome sets, this plant would likely be sterile because its three chromosome sets could not pair up properly during meiosis.

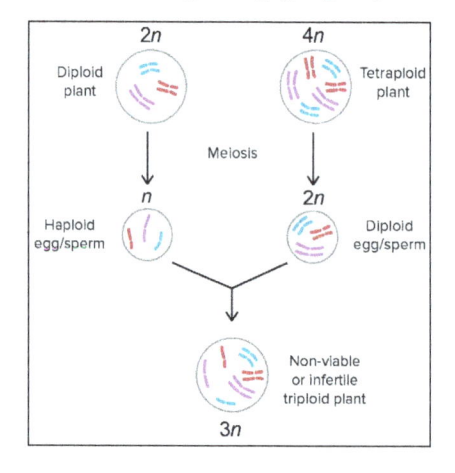

Because the tetraploid plants and the diploid species from which they came cannot produce fertile offspring together, we consider them two separate species. This means that speciation occurred after just a single generation.

Speciation by polyploidy is common in plants but rare in animals. In general, animal species are much less likely to tolerate changes in ploidy. For instance, human embryos that are triploid or tetraploid are non-viable—they cannot survive.

Sympatric Speciation without Polyploidy

Can sympatric speciation, speciation without geographical separation, occur by mechanisms other than polyploidy? There's some debate about how important or common a mechanism it is, but the answer appears to be yes, at least in some cases. For instance, sympatric speciation may take place when subgroups in a population use different habitats or resources, even though those habitats or resources are in the same geographical area.

One classic example is the North American apple maggot fly. As the name suggests, North American apple maggot flies, like the one pictured below, can feed and mate on apple trees. The original host plant of these flies, however, was the hawthorn tree. It was only when European settlers introduced apple trees about 200 years ago that some flies in the population started to exploit apples as a food source instead.

Rhagoletis pomonella.

The flies that were born in apples tended to feed on apples and mate with other flies on apples, while the flies born on hawthorns tended to similarly stick with hawthorns. In this way, the population was effectively divided into two groups with limited gene flow between them, even though there was no reason an apple fly couldn't go over to a hawthorne tree, or vice versa.

Over time, the population diverged into two genetically distinct groups with adaptations, features arising by natural selection, that were specific for apple and hawthorne fruits. For instance, the apple and hawthorne flies emerge at different times of year, and this genetically specified difference synchronizes them with the emergence date of the fruit on which they live.

Some interbreeding still occurs between the apple-specialized flies and the hawthorne-specialized flies, so they are not yet separate species. However, many scientists think this is a case of sympatric speciation in progress.

Coevolution

Coevolution refers to the evolution of at least two species, which occurs in a mutually dependent manner. Coevolution was first described in the context of insects and flowering plants, and has since been applied to major evolutionary events, including sexual reproduction, infectious disease, and ecological communities. Coevolution functions by reciprocal selective pressures on two or more species, analogous to an arms race in an attempt to outcompete each other. Classic examples include predator-prey, host-parasite, and other competitive relationships between species. While the process of coevolution generally only involves two species, multiple species can be involved. Moreover, coevolution also results in adaptations for mutual benefit. An example is the coevolution of flowering plants and associated pollinators (e.g., bees, birds, and other insect species).

Coevolution Examples

Predator-Prey Coevolution

The predator-prey relationship is one of the most common examples of coevolution. In this respect, there is a selective pressure on the prey to avoid capture and thus, the predator must evolve to become more effective hunters. In this manner, predator-prey coevolution is analogous to an evolutionary arms race and the development of specific adaptations, especially in prey species, to avoid or discourage predation.

Herbivores and Plants

Similar to the predator-prey relationship, another common example of coevolution is the relationship between herbivore species and the plants that they consume. One example is that of the lodgepole pine seeds, which both red squirrels and crossbills eat in various regions of the Rocky Mountains. Both herbivores have different tactics for extracting the seeds from the lodgepole pine cone; the squirrels will simply gnaw through the pine cone, whereas the crossbills have specialized mandibles for extracting the seeds. Thus, in regions where red squirrels are more prevalent, the lodgepole pine cones are denser, contain fewer seeds, and have thinner scales to prevent the squirrels from obtaining the seeds. However, in regions where crossbills are more prevalent, the cones are lighter and contain thick scales, so as to prevent the crossbills from accessing the seeds. Thus, the lodgepole pine is concurrently coevolving with both of these herbivore species.

Acacia Ants and Acacias

An example of coevolution that is not characteristic of an arms race, but one which provides a mutual benefit to both a plant species and insect is that of the acacia ants and acacia plants. In this relationship, the plant and ants have coevolved to have a symbiotic relationship in which the ants provide the plant with protection against other potentially damaging insects, as well as other plants which may compete for nutrients and sunlight. In return, the plant provides the ants with shelter and essential nutrients for the ants and their growing larvae.

Flowering Plants and Pollinators

Another example of beneficial coevolution is the relationship between flowering plants and the respective insect and bird species that pollinate them. In this respect, flowering plants and pollinators have developed co-adaptations that allow flowers to attract pollinators, and insects and birds have developed specialized adaptations for extracting nectar and pollen from the plants.

There are at least three traits that flowering plants have evolved to attract pollinators:

- Distinct visual cues: flowering plants have evolved bright colors, stripes, patterns, and colors specific to the pollinator. For example, flowering plants seeking to attract insect pollinators are typically blue an ultraviolet, whereas red and orange are designed to attract birds.

- Scent: flowering plants use scents as a means of instructing insects as to their location. Since scents become stronger closer to the plant, the insect is able to hone-in and land on that plant to extract its nectar.

- Some flowers use chemical and tactile means to mimic female insect species to attract the male species. For example, orchids secrete a chemical that is the same as the pheromones of bee and wasp species. When the male insect lands on the flower and attempts to copulate, the pollen is transferred to him.

Hummingbirds are another type of pollinator that have coevolved for mutual benefit. The hummingbirds serve as pollinators and the flowers supply the birds with nutrient-rich nectar. The flowering plants attract the hummingbirds with certain colors, the shape of the flower accommodates the bird's bill, and such flowers tend to bloom when hummingbirds are breeding. Coevolution of such flowering plants with various hummingbird species is evident by the distinct shape and length of the flower's corolla tubes, which have adapted to the shape and length of the hummingbird bill that pollenates that plant. The shape of the flower has also adapted such that the pollen becomes attached to a particular region of the bird while it consumes the nectar from the flower.

Adaptation

Adaptation is the process by which a species becomes fitted to its environment; it is the result of natural selection's acting upon heritable variation over several generations. Organisms are adapted to their environments in a great variety of ways: in their structure, physiology, and genetics, in their locomotion or dispersal, in their means of defense and attack, in their reproduction and development, and in other respects.

In biology this general idea has been coopted so that adaptation has three meanings. First, in a physiological sense, an animal or plant can adapt by adjusting to its immediate environment—for instance, by changing its temperature or metabolism with an increase in altitude. Second, and more commonly, the word adaptation refers either to the process of becoming adapted or to the features of organisms that promote reproductive success relative to other possible features. Here the process of adaptation is driven by genetic variations among individuals that become adapted to—that is, have greater success in—a specific environmental context. A classic example is shown by the melanistic (dark) phenotype of the peppered moth (Biston betularia), which increased in numbers in Britain following the Industrial Revolution as dark-coloured moths appeared cryptic against soot-darkened trees and escaped predation by birds. The process of adaptation occurs through an eventual change in the gene frequency relative to advantages conferred by a particular characteristic, as with the coloration of wings in the moths.

Against the background of a lichen-covered oak tree, a darkly pigmented peppered moth (Biston betularia) stands out, while the light gray moth (left) remains inconspicuous.

The third and more popular view of adaptation is in regard to the form of a feature that has evolved by natural selection for a specific function. Examples include the long necks of giraffes for feeding in the tops of trees, the streamlined bodies of aquatic fish and mammals, the light bones of flying birds and mammals, and the long daggerlike canine teeth of carnivores.

Adaptations The habitat adaptations of walruses (thick skin to protect against cold conditions), hippopotamuses (nostrils on the top of the snout), and ducks (webbed feet).

All biologists agree that organismal traits commonly reflect adaptations. However, much disagreement has arisen over the role of history and constraint in the appearance of traits as well as the best methodology for showing that a trait is truly an adaptation. A trait may be a function of history rather than adaptation. The so-called panda's thumb, or radial sesamoid bone, is a wrist bone that now functions as an opposable thumb, allowing giant pandas to grasp and manipulate bamboo stems with dexterity. The ancestors of giant pandas and all closely related species, such as black bears, raccoons, and red pandas, also have sesamoid bones, though the latter species do not feed on bamboo or use the bone for feeding behaviour. Therefore, this bone is not an adaptation for bamboo feeding.

Giant panda (Ailuropoda melanoleuca) feeding in a bamboo forest, Szechwan province, China.

The english naturalist charles darwin, in on the origin of species by means of natural selection, recognized the problem of determining whether a feature evolved for the function it currently serves.

The sutures of the skulls of young mammals have been advanced as a beautiful adaptation for aiding parturition [birth], and no doubt they facilitate, or may be indispensable for this act; but as sutures occur in the skulls of young birds and reptiles, which only have to escape from a broken egg, we may infer that this structure has arisen from the laws of growth, and has been taken advantage of in the parturition of the higher animals.

Thus, before explaining that a trait is an adaptation, it is necessary to identify whether it is also shown in ancestors and therefore may have evolved historically for different functions from those that it now serves.

Another problem in designating a trait as an adaptation is that the trait may be a necessary consequence, or constraint, of physics or chemistry. One of the most common forms of constraint involves the function of anatomical traits that differ in size. For example, canine teeth are larger in carnivores than in herbivores. This difference in size is often explained as an adaptation for predation. However, the size of canine teeth is also related to overall body size (such scaling is known as allometry), as shown by large carnivores such as leopards that have bigger canines than do small carnivores such as weasels. Thus, differences in many animal and plant characteristics, such as the sizes of young, duration of developmental periods (e.g., gestation, longevity), or patterns and sizes of tree leaves, are related to physical size constraints.

Adaptive explanations in biology are difficult to test because they include many traits and require different methodologies. Experimental approaches are important for showing that any small variability, as in many physiological or behavioral differences, is an adaptation. The most rigorous methods are those that combine experimental approaches with information from natural settings—for example, in showing that the beaks of different species of Galapagos finch are shaped differently because they are adapted to feed on seeds of different sizes.

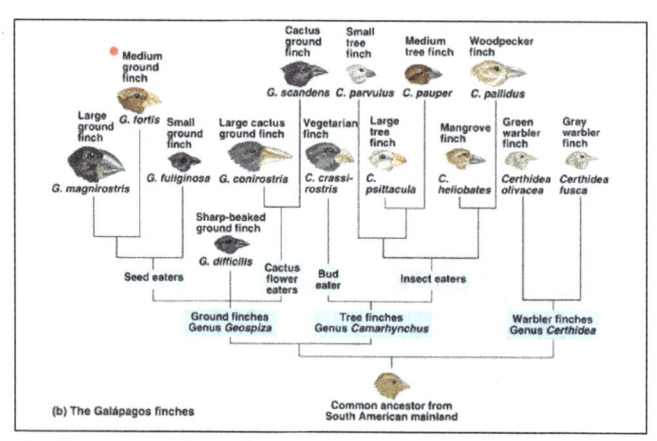

Fourteen species of Galapagos finches that evolved from a common ancestor. The different shapes of their bills, suited to different diets and habitats, show the process of adaptive radiation.

The comparative method, using comparisons across species that have evolved independently, is an effective means for studying historical and physical constraints. This approach involves using statistical methods to account for differences in size (allometry) and evolutionary trees (phylogenies) for tracing trait evolution among lineages.

Adaptive Radiation

Adaptive radiation refers to the adaptation (via genetic mutation) of an organism which enables it to successfully spread, or radiate, into other environments. Adaptive radiation leads to speciation and is only used to describe living organisms. Adaptive radiation can be opportunistic or forced through changes to natural habitats.

Adaptive Radiation Examples

Examples of adaptive radiation are all around us, in every living organism. No organism today is exactly the same as its original ancestor. Some species have changed significantly, such as the diversification from a single species into the elephant and the hyrax. One only has to look at the image below to understand how the selection of a different habitat or even a similar habitat but a different choice of diet can cause huge anatomical and physiological changes during the process of adaptive radiation.

Bush hyrax – the elephant's closest relative

Marsupials

One of the most common examples of the theory of adaptive radiation is the dispersion and diversification of the marsupials (metatherians) into different orders and species. Marsupials have developed from a single ancestor into multiple, diverse forms. This has happened in a continent completely cut off from the influence of many other species.

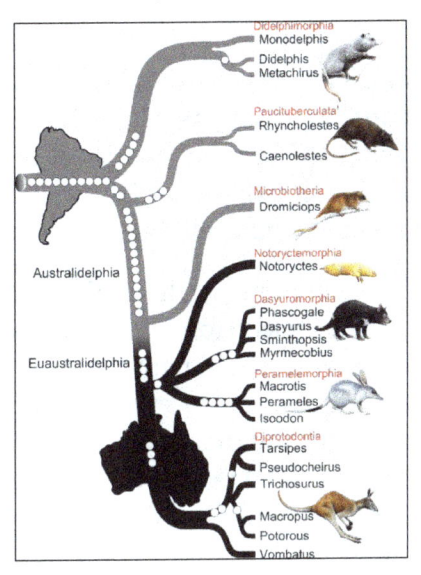

In the image above, seven orders of marsupials are shown with grey and black lines indicating South American and Australian habitats respectively. Yet each order has diversified from its superorder (Euaustralidelphia) through adaptation. Each order can better survive thanks to a specific adaptation to a different habitat.

This independent evolution in response to particular aspects of the environment is also mimicked across the globe by placental mammals. Many marsupials have developed in extremely similar ways to placental mammals living in similar environments, even though they have been cut off from these other populations since the breakup of the supercontinent known as Gondwana. This process has not yet ended. Today, Australia crawls to the north at a rate of about 3 centimeters per year.

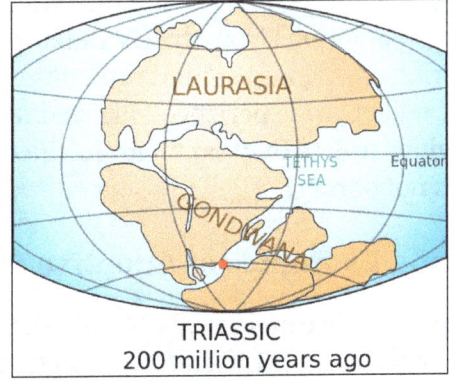

Early supercontinents – Gondwana and Laurasia

This separation of species, yet with similarities in both adaptations and environments, tells us that biodiversity is usually the result of adaptive radiation.

Skin Color

Human skin color is another example of adaptive radiation. The color of the skin is regulated by the presence of melanin, a natural pigment which in higher quantities can absorb ultra-violet light and protect the dermis. People with light complexions mainly produce pheomelanin which has a reddish-yellow hue, while those with dark-colored skin primarily produce eumelanin which is dark brown in color.

Under the rays of the sun, vitamin D synthesis is stimulated, while folate degrades. Folate is necessary for early fetal development and is partially regulated by UV exposure. Too little or too much sun can dysregulate folate levels. While the current theories of the human race originating from an African location are under discussion, using this model to explain adaptive radiation is helpful. In fact, this model can be used to explain two different types of adaptive radiation.

The first concerns very early ancestors of man (the hominids) who were largely covered with hair to keep them warm in largely forested areas. Hominid skin, protected by hair, was almost definitely not as dark as his early descendants. We do not have the fossil evidence to prove this, but mammals usually have much lighter skin when covered in thick layers of hair or fur, as opposed to mammals with thin coats. Upon migrating to more open savannahs where the hominids could hunt more successfully but directly under the rays of the equatorial sun, this hair became superfluous. To be protected from the UV rays of the sun they developed darker skin. This darker skin reduced the degradation of folic acid, meaning higher pregnancy and birth rates, while the constant availability of the equatorial sun meant that vitamin D production was sufficient to ensure good health.

When these populations eventually moved away from the heat of the equator and into colder regions high levels of melanin became more of a hindrance to the health and reproductive capacity of this migrating population. Skin did not need as much melanin to protect it from the meager sun; those with darker skin would block what little UV light there was and synthesize less vitamin D, leading to lower levels of health and fitness (rickets) and dysregulated folate levels (miscarriages).

Those who migrated to the far northern regions of the Arctic Circle became slightly lighter in color, but darker than would usually be expected according to this theory. This has been explained by their seafood diets which provide ample dietary vitamin D during the colder seasons, while a darker skin color protected these populations from the UV radiation of the sun further reflected by the snow-covered landscape during spring and summer months. Research today tells us that the female Inuit population are more likely to experience folic acid deficiencies than lighter-skinned females in colder, temperate regions unless they eat folate-fortified foods. This is perhaps the reason why the color of their skin is not darker.

Phylogenetics – Discovering Examples of Adaptive Radiation

Phylogenetic research into visible genetic traits and, later on, DNA sequences is far from new. Aristotle devised his Scala Naturae or Ladder of Life in the third century before Christ, splitting animals into two very basic (and very wrong) main groups – those with red blood and those without. This idea expanded over the centuries due to the many distinctive characteristics of non-related species which live in similar environments.

Phylogenetics is the study of the evolutionary steps a species has taken during the process of speciation. These steps lead to the creation of a phylogenetic tree, an extremely simplified version of which is pictured below. These trees can be rooted or unrooted, meaning coming from a single known original ancestor or from an unknown ancestor or group of ancestors respectively. Phylogenetic trees depict the evolutionary history of one or more species in relation to its ancestors.

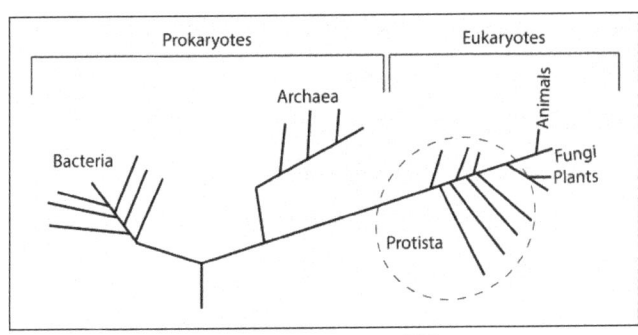

Phylogenetic tree of life

Ecological Opportunity – Making the most of Every Available Habitat

What has not yet been mentioned is that the term 'adaptive' in the context of adaptive radiation must indicate a move towards a healthier and more successfully reproductive species. While it is often understood that any evolution requires thousands if not tens of thousands of years to lead to a phenotype which is common to a group of organisms but not to the original ancestors, and due to changes in the environment, this can actually be a fairly rapid shift.

In order to move through the process of adaptive radiation, a population must nearly always be exposed to ecological opportunity. This ecological opportunity must be present in order for speciation can occur. The most important ecological opportunity as far as mammals are concerned was the mass extinction of the dinosaur, where both warm- and cold-blooded species could move into fresh ecosystems previously too unsafe or heavily populated.

The move from this ecological opportunity to the adaptive radiation of a population requires a complete set of traits which allow a species to take advantage of the new environment, such as herbivorous mammals migrating into a new, plant-filled ecosystem. This set of traits is referred to as a key innovation. The next step is ecological release – the proliferation of a population in a new environment without limiting factors such as competition or overpopulation.

Adaptive Radiation in Urban Environments – A Recent but Rapid Development

Urban environments, where ecosystems are very different from rural environments, are already bringing forth common genetic mutations in various plants and animals. Serotonin transporter gene (SERT) mutations in urban birds reduce levels of anxiety. This in itself is not observable in the anatomy of the bird, yet this mutation is associated with health- and survival-related traits such as physiological preparation for egg-laying and hatching success, with a subsequent increase in reproduction, and is therefore compliant with the laws of adaptive radiation.

Abiotic barriers, such as high heavy metal content in soil or water, can cause mutations in some species of plants which increase flavonoid synthesis, as higher flavonoid content increases heavy metal tolerance. Seed dispersal in urban plants can also be different from that of the same plants in other, less populated, polluted or protected ecosystems. Biotic variables have been previously believed to be more responsible for adaptive radiation than abiotic, but both can work together.

References

- Evolution-scientific-theory, science: britannica.com, Retrieved 13 March, 2019
- Convergence: utexas.edu, Retrieved 3 May, 2019
- Divergent-evolution: biologydictionary.net, Retrieved 11 January, 2019
- Parallel-evolution: biology-online.org, Retrieved 4 June, 2019
- Main-theories-of-evolution, biology: yourarticlelibrary.com, Retrieved 3 March, 2019
- Common-descent: biologydictionary.net, Retrieved 2 May, 2019
- Species-speciation, tree-of-life, biology, science: khanacademy.org, Retrieved 21 April, 2019
- Coevolution: biologydictionary.net, Retrieved 15 February, 2019
- Adaptation-biology-and-physiology, science: britannica.com, Retrieved 5 April, 2019
- Adaptive-radiation: biologydictionary.net, Retrieved 30 August, 2019

Permissions

All chapters in this book are published with permission under the Creative Commons Attribution Share Alike License or equivalent. Every chapter published in this book has been scrutinized by our experts. Their significance has been extensively debated. The topics covered herein carry significant information for a comprehensive understanding. They may even be implemented as practical applications or may be referred to as a beginning point for further studies.

We would like to thank the editorial team for lending their expertise to make the book truly unique. They have played a crucial role in the development of this book. Without their invaluable contributions this book wouldn't have been possible. They have made vital efforts to compile up to date information on the varied aspects of this subject to make this book a valuable addition to the collection of many professionals and students.

This book was conceptualized with the vision of imparting up-to-date and integrated information in this field. To ensure the same, a matchless editorial board was set up. Every individual on the board went through rigorous rounds of assessment to prove their worth. After which they invested a large part of their time researching and compiling the most relevant data for our readers.

The editorial board has been involved in producing this book since its inception. They have spent rigorous hours researching and exploring the diverse topics which have resulted in the successful publishing of this book. They have passed on their knowledge of decades through this book. To expedite this challenging task, the publisher supported the team at every step. A small team of assistant editors was also appointed to further simplify the editing procedure and attain best results for the readers.

Apart from the editorial board, the designing team has also invested a significant amount of their time in understanding the subject and creating the most relevant covers. They scrutinized every image to scout for the most suitable representation of the subject and create an appropriate cover for the book.

The publishing team has been an ardent support to the editorial, designing and production team. Their endless efforts to recruit the best for this project, has resulted in the accomplishment of this book. They are a veteran in the field of academics and their pool of knowledge is as vast as their experience in printing. Their expertise and guidance has proved useful at every step. Their uncompromising quality standards have made this book an exceptional effort. Their encouragement from time to time has been an inspiration for everyone.

The publisher and the editorial board hope that this book will prove to be a valuable piece of knowledge for students, practitioners and scholars across the globe.

Index

A
Adaptive Radiation, 171, 183, 210-214
Agamenogenesis, 102
Allopatric Speciation, 182, 184, 202-203
Angiosperms, 104, 107, 111, 142, 188
Archaeal Cells, 71, 80
Asexual Reproduction, 66, 69, 83, 87, 92, 95-96, 99-102, 107-108, 143

B
Binary Fission, 79, 82, 87, 93-96, 101, 191
Biological Inheritance, 143-144
Biomolecules, 77

C
Cell Cycle, 62, 75, 88-91
Cell Differentiation, 63-66, 68, 128, 137
Cell Division, 1, 17-18, 23, 42-43, 55, 57-62, 66, 69, 76-77, 87, 90, 93, 107, 135-136, 152-153, 203
Cell Wall, 36, 40, 55, 60, 72, 77-82, 88, 90, 93-94
Cellular Reproduction, 16-17, 87
Cellular Respiration, 16, 83-85, 87, 136
Central Vacuole, 72, 77, 79
Centriole, 77, 135
Centrosome, 57, 60, 77, 89
Chloroplasts, 17, 20, 22, 39, 41, 48-49, 54-55, 71-72, 77-79
Chromatid, 42-43, 60
Chromatin, 42-44, 48, 59, 65-66, 73-74
Chromosome, 9, 42-43, 59-61, 74, 78, 80, 87-88, 93-95, 105, 143, 153-162, 164-165, 195-197, 200, 203-204
Coenzyme A, 50, 84, 86
Coevolution, 166, 205-207, 214
Convergent Evolution, 173, 175, 177, 179-181, 185
Cortical Reaction, 135
Cytokinesis, 59-60, 88-90
Cytoplasm, 17, 20-22, 27, 35, 38-42, 46-48, 54-55, 57, 60, 65, 67, 71, 73-76, 78-81, 84, 88-90, 93-95, 129-130

D
Deoxyribonucleic Acid, 2, 7, 18, 41, 92
Dihybrid Cross, 148-151
Diploid Species, 164, 204
Diploid Stage, 97, 105
Divergent Evolution, 182-184

Dna Double Helix, 20, 42, 59, 66
Dna Synthesis, 61-62
Ductus Deferens, 113-114, 117-120

E
Embryogenesis, 104, 136, 140, 142
Embryonic Stem Cells, 4
Endoplasmic Reticulum, 17, 22, 27, 35-36, 38-39, 44, 71, 73, 75
Eukaryotes, 7-8, 10, 17, 55, 58-61, 71, 74-75, 82, 93, 101, 175

F
Flagella, 34, 57, 80-82

G
Gametogenesis, 104, 128, 142, 147, 158
Genetic Material, 1, 5, 9, 18-19, 21, 44, 54, 59, 62, 80-81, 88, 102-104, 151, 158-159, 163, 197
Genetic Mutations, 102, 213
Golgi Apparatus, 8, 17, 22, 27, 35, 38, 40, 71

H
Haploid, 88, 94-95, 97, 99, 104-107, 111, 128-130, 133, 135, 154, 159-160, 204
Homeostasis, 4, 6-7

K
Krebs Cycle, 84-86

L
Lamarckism, 185-189, 196
Luteinizing Hormone, 118, 127, 131-132
Lysosomes, 17, 22, 34-38, 41, 72, 77-78

M
Meiosis, 60-61, 83, 88, 94-95, 97, 102, 104-107, 128-130, 148, 150, 152-160, 162, 165, 204
Microvilli, 56-57, 72-73
Mitochondria, 17, 20, 22, 27, 36, 39, 41, 48-50, 54-55, 71, 75-76, 78-79, 84
Mitosis, 57-61, 69, 82-83, 88-90, 94, 105-106, 129-130, 152-153
Mitotic Spindle, 57, 60, 89, 135
Monohybrid Cross, 148-150
Multicellular Organisms, 3-5, 8, 16, 41, 58, 72, 92, 95, 102, 104, 106-107

Index

Multiple Fission, 94-95
Mutation, 2, 9, 62, 99, 101-102, 107, 143, 162-165, 185, 189, 194-196, 210, 213
Mutation Theory of Evolution, 194, 196

N
Natural Selection, 5-6, 9, 71, 98-99, 162, 164, 166-168, 173-174, 180, 182-183, 185, 189-191, 193-194, 196, 198-199, 203, 205, 207-209
Nucleosome, 43
Nucleus, 7-8, 11, 16-17, 20-22, 35-36, 39, 41-42, 44-48, 50, 55, 60, 64-65, 71, 73-76, 79-80, 88-89, 93-97, 105, 111-112, 118, 135, 154, 164, 172

P
Parthenogenesis, 96, 102-103
Peroxisomes, 35-37, 39, 41, 76
Phenotype, 143-144, 147, 149, 151-152, 159, 161, 182, 198, 208, 213
Phospholipid Bilayer, 26, 36-37, 40, 72
Plasma Membrane, 16, 23, 34, 38, 72-73, 75-76, 78, 81, 90, 135

Plastids, 3, 72, 77
Polyploidy, 160, 165, 197, 203-204
Primates, 169, 183-184
Prokaryotic Cell, 79-81

R
Ribonucleic Acid, 18, 42, 93
Ribosomes, 21-22, 27, 38-39, 41, 44, 46, 55, 74-76, 78, 80-81

S
Single-celled Organisms, 4-5, 7, 10, 16, 69, 95, 99, 102
Speciation, 166, 182, 184, 186, 191, 193-194, 199-200, 202-205, 210, 213-214
Spermatogenesis, 128-130
Spermatozoa, 113, 115, 118, 128-129, 157-158
Sporophyte, 97, 106-107
Sympatric Speciation, 182, 203-205

Z
Zygote, 70, 88, 91, 95, 99, 103-106, 110-112, 133, 135-137, 189, 204

CPSIA information can be obtained
at www.ICGtesting.com
Printed in the USA
BVHW011328100920
588441BV00003B/33